Post-Apocalyptic Environmentalism

Carl Cassegård • Håkan Thörn

Post-Apocalyptic Environmentalism

The Green Movement in Times of Catastrophe

Carl Cassegård
Sociology and Work Science
University of Gothenburg
Göteborg, Sweden

Håkan Thörn
Sociology and Work Science
University of Gothenburg
Göteborg, Sweden

ISBN 978-3-031-13202-5   ISBN 978-3-031-13203-2   (eBook)
https://doi.org/10.1007/978-3-031-13203-2

This Palgrave Macmillan imprint is published by the registered company Springer Nature Switzerland AG.
The registered company address is: Gewerbestrasse 11, 6330 Cham, Switzerland

PREFACE

This book springs from a twofold puzzlement. With each year, climate-related catastrophes seem to increase in frequency. Intense heat waves with extreme environmental consequences ranging from drought to forest fires have become commonplace. These events highlight that climate-related catastrophes are not just a future threat but happen here and now, and that they are becoming the new normal. Rather than *facing* catastrophe, we seem already to be in the *midst* of one—a catastrophe which, furthermore, the present system appears fundamentally incapable of stopping. Our first puzzlement is simply this: what are the implications of this for environmental activism? The environmental movement has long been dominated on the one hand by apocalyptic warnings of future threats and on the other hand by techno-optimistic reassurances that these threats can still be averted. Fear and hope have helped the movement mobilize, but in both cases the focus has been on the future. What happens to the movement—including its visions, strategies and motivating emotions—and how does it respond when the catastrophes that it has long warned about are already a horrible reality for a huge number of people and non-human life forms?

We are not only scholars, but also inhabitants of a burning planet. Our second puzzlement is therefore: what *can* be done in this situation and what remains meaningful to fight for despite the catastrophes? Here we find a second reason to turn to movements. As we hope to show, ample examples exist in today's environmental activism of political action that remains meaningful even in a time of catastrophes—not least struggling for justice and making practical efforts to lessen suffering.

In our book we expand on arguments that we first made in a 2018 article, 'Toward a postapocalyptic environmentalism? Responses to loss and visions of the future in climate activism' (*Environment and Planning E: Nature and Space*, 1(4), 561–578). We also draw on historical research conducted in connection with writing a book in Swedish on the history of the environmental movement to be published in 2023 (under the title *I skuggan av apokalypsen. Miljörörelser och industrikapitalism 1870–2020*). We also build on research on the environmental movement in which we have been engaged through various projects for more than a decade, as well as on our experiences as critical participants in the movement. We wish to thank, first of all, the activists who have shared their time with us, responding to interviews and allowing us to participate in their activities. Without their generosity, this book would clearly have been impossible. We also thank our colleagues who have contributed feedback at various stages in our writing, above all Alireza Behtoui, Frida Buhre, Sara Dahlman, Kristoffer Ekberg, Karolina Enquist-Källgren, Jennifer Hadden, Johan Hedrén, Emil Husted, Jonathan Josefsson, Anna Kaijser, Jochen Kleres, Björn-Ola Linnér, Eva Lövbrand, Sherilyn MacGregor, Karl Malmqvist, Joost de Moor, Johan Söderberg, Peter Strandbrink, Linda Soneryd, Sebastian Svenberg, Göran Sundqvist, David Slater, Adrienne Sörbom, Cyprien Tasset, Adam Tomkins, Ewoud Vandepitte, Mattias Wahlström, Carl Wilén, Magnus Wennerhag, Åsa Wettergren, and Özge Yaka. A special thanks goes to Åsa for her early encouragement of our ideas. We also thank the participants at conferences, workshops and seminars where we have presented portions of our research: CSM-RESIST Seminar at the University of Gothenburg, the 'Green Room' seminar at Linköping University 2018, the Centre for the Study of Political Organization at Södertörn University in 2018, the Global Studies seminar at Sophia University in 2019, the Environmental Human Sciences Seminar at Stockholm University in 2020, the 'Transformative Potential of New Environmentalism and Its Narratives' session at the *ESA (Europ*ean Sociological Association) conference in 2021, the 'Organizing for Apocalypse' workshop in Copenhagen in 2021, and the Higher Seminar at the Department of Sociology, Uppsala University in 2022. We are also grateful to the Swedish Research Council and Formas for financing two research projects ('The Environmental Movement in a Globalizing World:

Institutionalization and/or Transformation' 2013–2016, and 'Adapting to the Climate: Emotions and Narratives in the Environmental Movement', 2020–2023), and to Magnus Bergvalls stiftelse and Åke Wibergs stiftelse for providing additional funding for our two recent book projects.

Göteborg, Sweden

Carl Cassegård
Håkan Thörn

CONTENTS

1   Narrative and Nature Interest   1

2   Green Progress   29

3   Apocalypse   53

4   Postapocalypse   77

5   Towards a Critique of the Environmental Movement   113

Index   133

About the Authors

Carl Cassegård is Professor of Sociology at the University of Gothenburg, Sweden. His current research interests include the environmental movement, critical theory and eco-Marxism, and his books include *Toward a Critical Theory of Nature: Capital, Ecology, and Dialectics* (2021), *Climate Action in a Globalizing World* (co-edited 2017), and *Youth Movements, Trauma and Alternative Space in Contemporary Japan* (2014).

Håkan Thörn is Professor of Sociology at the University of Gothenburg, Sweden. His research concerns social movements, power, and globalization and his books include *Anti-Apartheid and the Emergence of a Global Civil Society* (Second Edition 2009), *Climate Action in a Globalizing World* (co-edited 2017), *Contemporary Co-Housing in Europe* (co-edited 2020) and *Urban Uprisings* (co-edited 2016).

CHAPTER 1

Narrative and Nature Interest

Abstract This chapter explains the book's purpose and theoretical framework. Ever since its emergence, the environmental movement has developed in tandem with the growth of capitalism. This forms the background for the book's guiding question: how is it possible that the catastrophic trajectory of the system has changed so little despite over a century of environmental protests? We search for clues in the history of the environmental movement, focusing on three overarching narratives: the narrative of green progress, the apocalyptic narrative and the postapocalyptic narrative—devoting particular attention to the postapocalyptic narrative, which is least well explored.. We also pay attention to their latent contradictions, which are linked to the different interests, which we seek to capture through the notion of nature interests.

Keywords Environmental movement • Narrative • Utopia • Ideology • Collective identity

A clear relation exists between industrial capitalism, environmental devastation, and environmental movements. Ever since its emergence in the nineteenth century, the environmental movement has developed in tandem with the growth of capitalism and its increasingly catastrophic

C. Cassegård, H. Thörn, *Post-Apocalyptic Environmentalism*, https://doi.org/10.1007/978-3-031-13203-2_1

'domination of nature'. The catastrophic character of this domination is registered in the anxiety, alarm, and terror inspired by the notion of the Anthropocene, the age in which humanity is supposed to be in charge yet in which many people feel painfully helpless in the face of a system that appears unstoppable (Cassegård, 2021, p. 7, Hamilton, 2017, p. viif, Stoner & Melathopoulos, 2015). The sense that the catastrophe is already here is also fuelling new forms of environmentalism. Among the significant strands of environmentalism today, we find not only the eco-modernist variety, with its sanguine belief in 'green progress', and its seeming opposite, the apocalyptic environmentalism that warns of future cataclysms that must be averted at all costs. We also find a growing presence of what we term a postapocalyptic environmentalism, which accepts that the apocalypse is already ongoing or unavoidable and sees such an acceptance as the premise for political action.

This forms the background for the guiding question of this book: how is it possible that so little is being done? Why has the catastrophic trajectory of the system been so little affected, despite more than a century of environmental protests? A follow-up question is: what *can* be done today to change this trajectory or at the very least to face ongoing or inevitable catastrophes in a way that minimizes suffering? In this book, we search for clues to how we might answer these questions in the history of the environmental movement. As Alberto Melucci (1989) points out, movements are 'prophets of the present', embodiments of the 'feelers' of possibility that all societies stretch out into the future. While to some extent all movements are enmeshed in the system and coloured by their times, they are also incubators of futures, generators of knowledge and new visions and ideas that might guide us.

For a long time, the environmental movement's rallying cry has been that we must act before it is too late. Even when protests have been directed at ongoing examples of environmental destruction, the tendency has been to stress the future disasters that will result if business goes on as usual. Arguably, since the beginning of the post-war era, environmentalism has been infused with a strong current of apocalyptic sentiment, finding expression in what we call an apocalyptic narrative. This narrative distinguished the 'new' post-war environmental movement from earlier forms of conservationist activism—and from much social movement activism in general—by its invocation of impending global doom as a tool to rouse action and mobilize support. Beginning with Rachel Carson's *Silent Spring* and other classics of environmentalist literature, the post-war environmental movement has typically achieved its greatest impact by invoking the terrifying losses that will ensue in the future unless we act to prevent them (e.g. Killingsworth & Palmer, 1996). This means that the environmental movement has often stood out when compared to other movements through its 'future-oriented pessimism'; dreams of a better or utopian future have been less important as a mobilizing tool than fear of the coming catastrophe or collapse (Thörn, 2019). Apocalyptic images of future catastrophes still dominate much of environmentalist discourse. Melting polar caps, droughts, hurricanes, floods, and growing chaos are familiar tropes regularly invoked by activists as well as establishment figures like Al Gore. As 'green' concerns have become prioritized policy issues in natural and global governance, these apocalyptic tropes have virtually become *comme il faut* within respectable discourse, losing much of their oppositional and system-critical edge in the process (Methmann, 2013).

At the same time, this apocalyptic narrative is being challenged in various ways within the environmental movement. First, in the context of established and highly institutionalized environmental movement organizations such as Greenpeace and WWF, the apocalyptic narrative has been criticized on the basis of strategic concerns. Here it is argued that messages emphasizing the hope of a possible better future should replace images of the apocalypse because the latter do not have a mobilizing function, but rather enforce a passive attitude among citizens, or, in the terminology that we use in this book, depoliticization. By depoliticization we mean a discursive process that removes issues from the sphere of political action, by naturalizing them, covering up conflict lines or presenting them as beyond the limits of human power. In contrast to the future-oriented pessimism characteristic of the apocalyptic narrative, these organizations therefore consciously adopt a strategy of future-oriented optimism.

Second, an entirely different anti-apocalyptic position, and attempt at politicization, is taken in the waves of protest that have infused fresh anti-institutional energy in the environmental movement in recent years. These protests seem to be nourished neither by a strong sense of hope, nor by fears of a future disaster, but by a sense and an idea that the catastrophe is already ongoing. We see it in the Global South, where activists are killed, communities destroyed, and populations displaced because of environmental destruction. We see it among the nuclear refugees of post-Fukushima Japan and the frontline communities of the Global North, as well as in the transition movement and in the cultural activism associated with groups such as Dark Mountain. We also see it in the speeches of

Greta Thunberg (2019), the iconic spokesperson of the largest protest wave of globally coordinated environmental protest so far (de Moor & Wahlström, 2019). In the wider discourse on the environment as well, there appears to be an increasing preoccupation with catastrophe, not so much as a future threat but as a present reality or as a future that can no longer be averted.

Next to the apocalyptic narrative and the resurgent future-oriented optimism, we are thus seeing many examples of what we refer to as a postapocalyptic narrative in which the catastrophe is experienced or imagined as 'already here', to use Erik Swyngedouw's (2013, p. 15) expression. By postapocalyptic environmentalism we mean environmental activism based on catastrophic losses experienced as already having occurred, as ongoing or as impossible to prevent, rather than as a future risk or threat. The term postapocalyptic may seem paradoxical since it implies that life 'goes on' even after the end. We find it useful, however, as a contrast to the apocalyptic quality of much environmentalism and to highlight affinities with postapocalyptic themes and sentiments in contemporary culture (see Berger, 1999; Williams, 2011; Kaplan, 2016). It is an environmentalism which, we suggest, is increasingly prevalent, especially among groups that are radical in the sense that they lack trust in the established institutions governing the environment and in the capacity of capitalism to reform itself into a green, just, and 'sustainable' form.

In the book, we chart the history of the environmental movement, in the nineteenth and twentieth centuries, but focus primarily on recent and contemporary trends with a clear bearing on the present situation. Our focus is on three overarching narratives, or what we call 'meta-narratives', within the environmental movement: the narrative of green progress; the apocalyptic narrative; and the postapocalyptic narrative. Among these, we devote particular attention to the postapocalyptic narrative, which is the least well-explored and has only gained a prominent presence in the environmentalism of the Global North in recent years. Above all, we seek to explore and clarify the notion of postapocalyptic politics by focusing on how postapocalyptic narratives can be deployed in political mobilizations. Rather than necessarily implying passivity or mere mourning, we contend that postapocalyptic narratives can be the wellspring of a *postapocalyptic politics* in which activism arises as a response to loss. This politics, however, can only be understood by scrutinizing the particular utopian imagination brought into play by this narrative, its relation to time-space and the way it constructs collective identity. In order to discern the senses in which the postapocalyptic narrative is both rooted in, and different from, apocalyptic imagination, and how both of these discourses are radically different from the social movement narratives that defined Enlightenment modernity, including that of green progress, we will also revisit the origins of environmentalist thinking in the nineteenth and twentieth centuries.

We discern three periods in the history of the environmental movement. First, in the period from the late nineteenth century up until the Second World War, in a global conext dominated by the nature conservation movement, the narrative of green progress rises to dominance in the environmental movement. Second, with the rise of the ecological environmental movement in the post-war period, the apocalyptic narrative becomes dominant in environmentalism across the world. Third, beginning in the early 2000s, the climate movement starts to come to the fore in environmentalism globally, boosting the postapocalyptic narrative. Importantly, we do not see this as a linear development, nor do we imply that one narrative replaces another. On the contrary, we argue that the three meta-narratives are an integral part of the articulation of experiences of environmental destruction brought about by industrial capitalism. Consequently, we will demonstrate how the three narratives are present in different parts of the environmental movement throughout the three periods. In the early twentieth century, the narrative of green progress was being challenged by apocalyptic as well as postapocalyptic narratives articulated in the margins of the movement. In the decades after the Second World War, the dominant apocalyptic narrative co-existed with new and influential articulations of green progress, while the postapocalyptic narrative was still marginal. Finally, in the present period, apocalyptic imagery still abounds in environmentalist discourses, while a redefined narrative of green progress has gained substantial support, e.g. with the Green New Deal. In the early twenty-first century, we find it difficult to argue that any of the three narratives has risen to a dominant position. A novelty with twenty-first-century postapocalyptic environmentalism, however, is that activism is embedded in a larger, global discourse in which the decisive catastrophe is emphatically presented as ongoing or unstoppable.

In the following sections, we will introduce our conceptualization of the environmental movement and of movement meta-narratives. Based on this analytic framework, we will formulate a set of questions that will guide our analysis of narratives in the chapters that follow, linking it to the concepts of *ideology* and *utopia*. First, we will explain how we conceive of the environmental movement both in terms of unity and difference, as a movement of movements. We will then discuss the relation between movement narratives, concepts of nature, and social structure in the context of the environmental movement, by introducing our concept of *nature interest*.

THE ENVIRONMENTAL MOVEMENT AS A MOVEMENT OF MOVEMENTS

Seen in a global perspective, the environmental movement is an extremely complex, heterogeneous, and even contradictory phenomenon. As such, does it really make sense to argue that there is *one* environmental movement? Our answer is yes, on the condition that we understand it in terms of a movement of movements (Klein, 2000). All major social movements are heterogeneous and contradictory and this is a fundamental part of their social dynamic. Nevertheless, the usefulness of the concept of 'social movement' has itself been questioned on these grounds. In the 2000s, *contentious politics* (McAdam et al., 2001) was influential as an analytical strategy to analyse collective protest. With a more limited analytical focus on strategic, confrontative action that targets the state, all the difficulties associated with capturing the unifying elements of a social movement are avoided. However, we argue that the concept of social movement is needed to capture collective action that is sustained over long periods, and with a strong continuity. In this book, we will show how there are strong elements of continuity in the 150-year-long history of the environmental movement. We also hope to demonstate how this history needs to be taken into account in order to fully understand the developments of this movement in the 2000s when it has repeatedly exhibited a global unity in large demonstrations at UN meetings and during climate strikes.

We use 'social movement' as an analytical concept (Melucci, 1989), constructed with the intention of providing analytical clarity to the complex realities of contemporary collective action. While an analysis of the complex reality of environmental activism needs to consider various forms of actions, interactions, text production, emotion rules, claims, and discursive positioning, the privileged element in our analysis is that of *collective identity*. As argued above, collective identity signifies the unity that is shown on the manifest level of environmental movement action—that is, in the shape of various collective actions, such as demonstrations, direct actions or manifestos. Collective identity is a name for the symbolic unity that holds a social movement together; it is a symbolic act produced through narratives. The questions we ask about movement narratives in this book address both their manifest and latent level, meaning that we intend to capture both how they manifest a collective identity, as well as the latent tensions and contradictions that reveal how they are anchored in particular economic, political, and cultural contexts. Thus, we will analyse how environmental movements emerge, how they are sustained and change by examining their narratives on the exploitation and desctruction of nature, who or what is the cause of this, what is to be done about it, who should do it, and how it should be done. Questions that guide our analysis also concern how the narrative articulates relationships between society and nature, spatial configurations, nature interests, and relationships between past, present, and future.

Ideology and Utopia in Social Movement Narratives

Collective identity is what ultimately holds the various and often heterogeneous elements of a social movement together; and collective identities are produced through narratives. Typically, such narratives invoke a collective 'we' that is called to act in the present, the action being made meaningful in the light of both past and future. This 'we' is not merely a cognitive but also an emotional construct: emotion generates identification with the movemnt, motivates action, and sustains commitment over time. Social movement narratives can be local or historically specific. For example, two different 'narratives of defeat' were constructed in the global environmental movement in the early 2010s; in connection with the 'post-Copenhagen depression' (following the failure to reach a substantial climate agreement at the UN climate meeting in 2009, Cassegård et al., 2017; de Moor & Wahlström, 2019); and in the US movement (in response to the lost political battle over the climate bill in 2009, Hadden, 2017). However, social movements also produce and sustain meta-narratives, which may assign meaning to a multiplicity of local or time-specific narratives by constructing the movement as an actor with a significant mission in human history. Such meta-narratives may both reflect and challenge hegemonic narratives in national or global contexts. Below, we refer to such meta-narratives simply as 'narratives' for short.

As *action-oriented* discursive constructions, social movement narratives always involve a certain form of temporality. They refer, on the one hand, to past developments that have created a problematic present, and on the other, to a future made possible by collective action that the narrative is designed to instigate. In this sense, social movement narratives always have a modern and a utopian element. They are *modern* in the sense that they are rooted in Enlightenment thinking, which postulated that humans have the capacity to create a world of their own making without the involvement of extra-social forces (be it God or Destiny). With the revolutions of the late eigtheenth and early nineteenth century, this secularized perspective on human action was embedded in collective action that aimed at creating a new kind of Man and Society. In this sense, these revolutions gave birth to the modern social movement as a secularized form of utopian collective action. Revolutions were also referred to as historical and material 'proof' of the ideals of the Enlightenment and modernity, and especially of the idea of human history as a movement of linear progress towards ever more developed forms of life. The narratives and symbolic power of these revolutions thus involved a re-articulation of two early modern utopian discourses in which the modern social movement meta-narrative is rooted: first, the utopian literary tradition initiated by Thomas More, with its ambivalent spatiality—utopia as a 'non-place' (*outopos*) and a 'desired place' (*eutopos*); and second, Christian Chiliasm (or Millennialism), with its ambivalent temporality—the notion of the coming of the Millennium that represented both the end of time and the beginning of a new era, in which the chosen few would enter a sacred social order defined by peace, justice, and plenty. In a sense, the narratives of the first modern revolutions implied not only a secularization of Chiliasm's promise of the coming of a New Time but also its marriage with utopia's ambivalent concept of space. Utopia as 'nowhere' and 'desired place' is articulated in the modern social movement narrative as a becoming through its location in the future (Thörn, 1997a).

Nineteenth- and twentieth-century social movements, defined by their forms of utopian action, thus presented variations of a narrative in which events in the past lead up to a problematic present, and in which collective action is needed to set society on the right path towards the future. These narratives were always defined by ambivalence regarding the role of agency in relation to historical movement. On the one hand, in line with the meta-narrative of human progress, historical movement towards a brighter future was inevitable. On the other, human agency was needed to give birth to the new, brighter future. In Marx, the proletarian revolution is articulated both in terms of a historical necessity, and as a future event that can only take place as a result of the formation and the agency of a working-class collective subject. Fanon (1987) gave the ambivalence a different expression through his anti-colonial rearticulation of this universalist narrative, rejecting the Western idea of historical progress/evolution as complicit with colonialization under capitalism, while retaining the idea of a future-oriented collective agency. By contrast, the narrative of green progress retained the idea of progress—objecting to the colonization of nature but only by advocating certain limits to capitalist expansion, which needed to be carefully controlled and steered in a way that avoided the total destuction of nature. Post-war apocalyptic environmentalism represents a sharper break with the future-oriented optimism of the social movements that had defined political modernity since the late eighteenth century by presenting a narrative designed to mobilize collective action not so much by promising a bright future as by the need to avoid disaster. As already mentioned, the recent crisis of the apocalyptic narrative may be seen as a response to its depoliticizing consequences, its failure to instigate mass collective action. The attempt by established environmental organizations to respond to this by a new politicizing discourse of hope has strong affinities with the Enlightenment legacy. By contrast, the postapocalyptic narrative has roots in the apocalyptic tradition while also representing a break with it. Does this mean that apocalyptic and postapocalyptic narratives are anti-utopian? As we will show, utopian elements are in fact strongly present in these narratives, although they are delinked from the idea of linear progress. No social movement, we suggest, can dispense with utopias.

Importantly, social movement narratives always involve both friends and foes; the collective lead character, the acting we, is defined by its boundaries in relation to enemies of the past and the present, and the allies which can assist in bringing about a better world. The moment of constructing a collective, integrated 'we', with well-defined interests and strategies, implies discursive closure, a fixation of meaning; it is the ideological dimension of a social movement narrative (Laclau, 1990, p. 92). At the same time, the polemical nature of this identity-construction often implies a certain ambiguity. As Jacques Rancière (1999) points out, a new political subject that comes into being seldom conforms entirely to established categories, and therefore often appears paradoxical and outrageous. An example of this, as we will see later, is when 'nature' is articulated as a subject of rights in the climate justice movement. The utopian dimension, as the element of future orientation, implies even more of this narrative openness and uncertainty. As an imagined future, the utopian element still also represents More's nonplace, an outside of the present society from which it is possible to reflect on, and critique, the existing social order. Using the formulation of Paul Ricoeur (1986, p. 15) in his *Lectures on Ideology and Utopia*, we thus argue that the utopian moment of the construction of a social movement narrative represents 'the ability to conceive an empty place from which to look at ourselves'.

Finally, the characters in social movement narratives are imagined actors situated not only in the movement of time but also in space—the immediate space in which collective action is supposed to take place (e.g. the local, national, or global space of action) as well as the desired space (local, national, or global utopia). As argued by Bakhtin (1981), time and space are not just fundamental structural elements in any narrative, but they are also inseparable; his concept of the chronotope (time–space) implies that a particular conception of time always has a spatial dimension and vice versa. Using the concept of the chronotope in the analysis of social movement narratives implies discerning how the collective that acts and moves—be it the nation, the tribe, or Humanity—always implies a particular ordering of time and space. While the identity, actions, and destiny of a given collective constitute the *content* of a meta-narrative, the temporal and spatial dimension is the *form* which organizes the narrative's content. In line with this, our analyses in this book will involve paying attention to both the intervowen temporal and spatial dimensions of concepts such as utopia and apocalypse.

In recent decades, we have seen a 'narrative turn' in social science (e.g. Czarniawska, 2004). In this context, collective narratives are most often ascribed the function of producing social cohesion. In the context of social movement studies, however, narratives are also analysed through their capacity to instigate collective mobilization and thereby contribute to social change (de Moor & Wahlström, 2019). Our approach differs from both of these perspectives in the sense that we approach social movement narratives with *a double perspective* (Thörn, 1997b). Drawing on Fredric Jameson's *The Political Unconscious* (Jameson, 1981), we treat social movement narratives as *symbolic acts*, which may involve (transformative) utopian energies as well as ideological functions. On the one hand, we want to analyse narratives in their own right as a kind of meaning-creating activity inherent in any collective effort to intervene in, and change, the social world. On the other hand, we will also analyse and deconstruct the ideological function of narratives, by making visible the conflicts and contradictions that they are designed to harmonize. This also involves making a distinction betweeen the manifest and the latent levels of the text (Thörn, 2015), which need to be seen as an articulation of the latent and the manifest levels of social movement action (Melucci, 1989). The *manifest level of social movement action* is defined by the processes through which it becomes a collective by performing as a unified actor in public space. This is where its collective actions are staged, where it addresses the public as well as its enemies and potential suppporters, articulates its identity, its purpose, its target. The public dimension of social movement action is visible in demonstrations, direct action, or in various movement publications (Thörn, 2015). This articulation of the movement's collective identity is a utopian act because it involves an imagined future; and it is ideological because it harmonizes the movement's internal, conflictive, and contradictory elements. These conflicts are visible on the *latent level of social movement action*, in the everyday practices in which various groups and individuals struggle to translate different social contexts and social conflicts into movement practice, and, ultimately, symbolic unity. In such processes of building collective unity out of heterogeneous elements—which may be social, material, and textual—narration is crucial. Such narration is produced through pamphlets, manifestos, and books written by movement intellectuals (Eyerman & Jamison, 1991) as well as by 'traditional' intellectuals whose works are adopted by the movement. Such texts form a particular genre, that of 'movements text' (Thörn, 1997b, 2019), which is the main empirical material used in the historical sections of this book.

We argue, and this is an important methodological strategy for us, that the latent level of conflict and contradiction leaves traces on the manifest level of movement narrative—in the shape of certain ambivalences around which the texts are organized both cognitively and emotionally. Such ambivalence can be read in terms of *attempts to reconcile or 'negotiate' the internal movement tensions, as well as the systemic contradictions and conflicts* that define the political spaces in which movements construct their action (Thörn & Svenberg, 2016).

The narrative of green progress, which involves a specific articulation of the above-mentioned ambivalence regarding the role of agency in relation to historical movement as progress, can be articulated with different emphases. For example, in the run-up to the 2009 Copenhagen UN climate meeting, the environmental movement tended to emphasize hope and *the necessity of collective agency* to push politicians into a binding climate agreement. In the wake of the failure to produce such an agreement, the movement, in its attempt to sustain mobilization amid a moment of defeat, emphasized fortitude and *historical necessity*: while the movement had lost an important battle, 'history was on their side'.

To sum up, our analytical double perspective on social movement narratives means that, on the one hand, we highlight the basic manifest features of these narratives, and on the other, we also pay attention to their latent tensions and conflicts. Such conflicts have often concerned how to conceptualize nature, its relation to society, and which forms of environmental action this relation implies.

Second Nature

Internal conflicts in the environmental movement have often centred on *the instrumental concept of nature*, emphasizing the utility of nature (see also our discussion of this further below). Critics have argued that the instrumental concept of nature must be abandoned altogether, and that human interaction with the environment should instead be guided by an ethics based on a holistic understanding of life on the planet. Previous research on the environmental movement has particularly highlighted two different concepts of nature that have been mobilized in opposition to the instrumental concept of nature. First, an *ethico-religious concept of nature* was prominent in early environmentalism, where it was used to emphasize a holistic understanding of the relation between society and nature. Such a concept of nature has also been articulated in the contemporary postapocalyptic narrative of indigenous environmental activism and the rights of nature movement.

Second, *a romantic concept of nature* emerged in reaction to the instrumental concept of nature. In the romantic concept of nature, nature is constructed not just as separate from, but as opposed to, society. Nature is here constructed as the Other in relation to civilized culture and humanity as such; it represents the wild and untamed as materialized in virgin forests, oceans, and mountains. At the same time, nature offers a possible source for humans to cultivate a 'true' relationship to their inner selves. Thus, the romantic concept of nature is mobilized in a critique of civilization in which a key theme is the alienation of humans from nature, and thus from their own nature.

The literature on the instrumental, ethical-religious, and romantic concepts of nature is vast, and this is not the place to discuss their many variations. Instead, this book will contribute to the understandig of how environmental movement narratives have played a part in establishing these concepts as well as modifying and criticizing them, and how this has always ocurred as part of the mobilization of collective action to address environmental destruction. More importantly, we also introduce a concept of nature that we call *second nature*. We argue that this concept has been a key part in thinking about, and acting on, the relation between society and nature, particularly in the ecological discourse on 'the environment' that was established after the Second World War. Nevertheless, it has received little attention in the research on the environmental movement and more broadly in the literature on concepts of nature and environment.

While the romantic concept of nature, in which nature is society's Other, was originally articulated in opposition to the instrumental concept of nature that rose to dominance with industrial capitalism, the concept of second nature was in turn constructed in opposition to the romantic concept of nature. Inspired by the Hungarian philosopher Georg Lukács (1971), many Marxists have used the term 'second nature' to describe the 'social laws of nature' established by capitalism. The starting point is Marx's idea that capitalist society appears as nature. The interaction between individuals, he writes in *Grundrisse*, gives rise to the appearance of 'an *alien* social power standing above them', a power that appears natural or as if it had arisen spontaneously from nature (Marx, 1973, p. 197). The result is that their conditions of existence 'appear as if they were *natural conditions*, not controllable by individuals' (ibid., p. 164). While, on the one hand, the expansion of capitalism on a planetary scale ensures that pristine nature hardly exists any longer, on the other hand, society itself starts to appear as nature.

Central to the theory of modern society as second nature is the insight that the dominion over nature attained in modern times not only liberates humankind, but also submits it to new forms of power since its own creations—economics, technology, and social conventions—turn against it in the guise of second nature. For example, the Marxist literary critic Walter Benjamin writes that 'technology confronts contemporary society as a second nature, and indeed, as economic crises and wars show, it is a nature no less elemental than that of ancient society' (1991, p. 444). The naturalization of society is reflected in the metaphors we use to describe the latter. Depictions of cities and urban life are a case in point. Benjamin (1997, p. 60ff) points out that the sea is Victor Hugo's model for the crowd, while the urban scholar Jane Jacobs (1992, p. 444f) suggests that urban environments 'are as natural as colonies of prairie dogs, or the beds of oysters'. Cities as a whole can appear as superhuman entities beyond human control, as can capitalist markets. Such things usurp the life that was meant for its human users, who are conversely reduced to components of the ecology of second nature.

How can we understand this semblance of nature in society? It is of course easy to dismiss it as a mere smokescreen that obscures reality, but such a dismissal would be premature. To a certain extent, what we think of as 'first' or 'real' nature is also deceptive—as, for example, when biological 'facts' are used to legitimize racism and patriarchy. A more rewarding approach, we believe, is to look at what makes things sometimes appear natural and sometimes as historical. Here, Benjamin's (1985) idea of 'natural history' offers a clue. In his study of the German baroque drama, he distinguishes between 'history' and 'nature' as two moments in how our environment manifests itself. Nature is what appears as original and given, while history appears as created and changeable by human action. Benjamin notes that the historical often petrifies into nature, into a frozen image of timelessness, while the natural conversely contains traces that reveal it to be perishable and historical. To one-sidedly focus on the process whereby history solidified into second nature would risk reinforcing the latter's ahistorical semblance of inevitability. Instead, nature itself should be shown to be historical and hence changeable. From this perspective, second nature is not necessarily falser than first nature, since both contain historical elements that have been obscured and rendered ahistorical.

This position can easily be developed into a full-blown constructivist position on nature. This step was taken in the 1980s by the Marxist geographer Neil Smith (2010) who argued that capitalism not only controls or exploits nature, but also produces it. On the basis of passages in *Grundrisse* and the first volume of *Capital*, Smith argues that all nature is in principle 'second nature' in the sense that, with few exceptions, it has already been drawn into and subordinated to capitalist exchange relations. This approach is close to the anti-essentialism associated with feminist and anti-racist critiques of how biology has been used to legitimize socially constructed hierarchies. As the philosopher Kate Soper (1995) observes, a tension exists between the concepts of nature used in ecologism and in feminism, with the former being 'nature-endorsing' and the latter 'nature-sceptical'. Our point, however, is not to advocate a one-sided constructivism. The intention of Marx, Lukács, and other thinkers who helped develop the idea of second nature was not to completely or unilaterally collapse nature into history, but rather to show how they dialectically condition and reshape each other. For the Frankfurt School philosopher and sociologist Theodor W. Adorno (1973), for example, it was in situations where the semblance of nature was strongest that a nature-sceptical or constructivist criticism of it was justified, while conversely nature should be *defended* where it seemed most threatened by historical or social processes. The degree to which a second nature has been produced by human action in recent decades has led some scholars to confuse it with 'the end of nature' (McKibben, 1989), the idea being that man's intervention into nature has altered it 'beyond recognition' (Wapner, 2010, p. 5). We argue that the idea of the end of nature builds on the assumption that there once existed a wilderness which has been erased by modern society. However, as we have already remarked, humans have altered nature in profound ways for tens of thousands of years. For sure, society's production and thus alteration of nature today is qualitatively different from pre-industrial times, but the idea of the end of nature is not helpful in understanding this new planetary condition. Instead, we argue that the concept of second nature brings more clarity to the current relationship between society and nature. For example, in this book the idea of second nature is relevant to understanding the development of the environmental movement, since it helps to explain how environmental activism can arise in relation to man-made environments. Significant strands of the early environmental movement arose in defence of cultivated landscapes that were perceived to be threatened by industrial capitalism, rather than in defence of wilderness. When the North American environmental justice movement denounced the latent racism that manifested itself in the fact that black people and the poor tended to suffer disproportionally from pollution, its activism revolved around second rather than first nature. The same holds true for urban environmentalism, including the protection of urban neighbourhoods, protests against bad air and other forms of pollution, and urban gardening. In general, the so-called environmentalism of the poor has been more about defending particular ways of life than any pristine, first nature. Second nature is central also in today's climate activism. Not only is its primary goal usually to protect human society rather than wilderness, but in addition the climate is already so affected by human activities that it can legitimately be considered second nature. These examples of activism show that the idea of second nature does not necessarily have a passifying or depoliticizing effect. It can itself become the terrain of new struggles that make nature appear changeable. Rather than simply dismissing second nature as false and the first as genuine, it may be more worthwhile to follow the lead of Benjamin's idea of natural history and try to trace how the historically created or artificial can appear as nature and how this nature in turn can become a breeding ground for historical action.

Second Nature as 'Environment', Ecology, and Anthropocene

While the concepts of 'environment' and 'ecology' were born in the context of natural science (Warde et al., 2018), the post-war environmental movement played an important role in the process through which these concepts became established in public discourse and everyday language (Jamison, 2001). This meant that the concept of nature that we call second nature became part of the everyday perception of the relation between society and nature. Both 'environment' and 'ecology' involve an implicit or explicit critique of the dualistic thinking that defines the instrumental and the romantic concepts of nature. The first of the 'ecological laws' defined by the influential natural scientist and environmental movement intellectual Barry Commoner (1971, p. 33) reads: 'Everything is connected to everything else. There is one ecosphere for all living organisms and what affects one, affects all'. In other words, human and natural history are inexctricably interlinked. While the distinction between society and nature is still significant in the environmental discourse that was established in the post-war era, it is a distinction perceived as a differentiation *within* a planetary ecosystem. Society and nature form parts of a whole that consists of different life forms with changing relations to each other. Thus, the conflict between society and nature remains in the environmental discourse's system perspective, whether it is articulated in terms of 'ecology' or 'earth system'. This means that even when it is recognized that modern society's second nature is a product—or an extension—of human activity, humans are still in the last instance a product—or an extension—of nature.

The concept of the Antropocene can be understood as arising when the earth as a whole comes to be understood as a second nature. Debates about the Anthropocene demonstrate how second nature can be articulated in different ways with regard to the extent to which humans can control nature. On the one hand, the concept of the Anthropocene has political implications: that humanity has gained the capacity to affect the geology, climate, and ecological system of the earth implies that it is possible to fundamentally transform existing planetary conditions through human action. On the other hand, we also see a tendency to depoliticize the notion of the Anthropocene by defining it purely in scientific terms. In this version, the Anthropocene is constructed as an objective fact, and humanity as a homogeneous and blind geological force beyond rational control, in the style of volcanic eruptions and variations in solar radiation. That this happens in the very era in which humanity's power to influence the earth's development is said to be greater than ever may appear paradoxical, but it is, we suggest, an expression of the naturalized or fetishistic form of social relations under capitalism. It reflects the fact that both society and nature have become subjected to the rule of impersonal forces that Lukács and Benjamin saw as characteristic of second nature.

Critics of the concept of the Anthropocene have suggested replacing it with the concept of the Capitalocene (Malm, 2016; Moore, 2016) to emphasize how climate change and other forms of environmental destruction is not the work of Humanity but the result of a class-driven process, namely the global expansion of capitalism. This perspective is eminently reasonable, considering the close connection between economic development and the use of energy—in particular fossil fuels—and other resources. Rather than reflecting a human constant, the destructive trends are rooted in the needs of a historically unique and abnormal system, namely capitalism, which depends on endless growth for its survival. This perspective emphasizes how environmental destruction can only be stopped by confronting this system through decisive political action, without which technical solutions will be in vain. We subscribe to this view and argue that politics always involves the articulation of conflicting interests, which are ultimately rooted in the social inequalities that define capitalist society. Should such conflicts not exist, political action to address environmental destruction would not be needed, as the issue could be solved by science, technology, and administration.

For us, such a political interpretation of the Anthropocene—in terms of the Capitalocene—does not necessarily involve an assumption that humans may transform the world as they wish through political action. The reality any contemporary political action needs to face is that nature is now striking back as a deadly force against human societies, ultimately threatening them with a fundamental collapse. This is the insight around which the postapocalyptic narrative is organized. But how does transformative political action then become possible in the light of such an insight? This is a key question in the present book and points to its central theme: that fundamental societal transformation as an important element involves narratives that assign an overarching meaning and a sense of historical direction to collective action.

NATURE INTEREST

Histories of environmentalism have often tended to focus too narrowly on concepts of nature, failing to fully grasp how environmental thinking is anchored in material and structural conditions. Through the concept of 'nature interest', we intend to capture and theorize the intersection between narratives and material interests, along the lines of a Marxist tradition that has emphasized 'praxis', as in 'cognitive praxis', a concept used by Eyerman and Jamison (1991) in their analyses of environmental movement history. Similar to Habermas's (1971) concept of 'knowledge interest', *nature interest* signifies a particular mode of constructing knowledge about nature. More than these authors, however, we want to stress how this knowledge production is embedded in structural societal conditions. The concept of nature interest is intended to capture how social movement practice conceptualizes the relation between society and nature (and how this relationship can be transformed by activism); and how this practice is embedded in the social structures that constitute its material conditions. Our concept of nature interest involves three dimensions:

First, nature interest refers to the production of knowledge and meaning, and in connection with this also that of emotions, about phenomena conceptualized as 'nature'. This may involve the articulation of a particular concept of nature, always defined by its relation to society or humanity. Examples include the romantic or ethico-religious concepts of nature discussed above. Instead of seeing nature as unilaterally determining action (as in the so-called reflexion theory found in socio-biology), we suggest a more dialectical and performative approach by understanding concepts of nature as produced through praxis. Concepts of nature and accompanying emotional relationships to nature emerge as a result of certain practices. Through the repetitions of such practices, concepts of nature can become institutionalized in a particular society, meaning that they may guide action in relation to nature in that society. Nevertheless, new practices may modify institutionalized concepts of nature, and assign them context-specific meanings.

Second, nature interest involves a strategic dimension, including the articulation of goals and the means for achieving them; and when these goals have been partly or fully achieved, the means for defending them.

For example, strategic environmental action may involve establishing a formal organization with the aim of defending nature by instituting nature reserves. And when that goal is partly achieved, as when a government passes a law on nature reserves, the defence of that law can also be of concern to the movement, signifying its institutionalization (in the sense of a movement's professionalization, formalization, and regularization of access to increasing cooperation with authorities, Thörn, 2022). This dimension of nature interest thus refers to strategic collective action in relation to nature, and how such action shapes the movement's collective identity.

Third, nature interest also includes a social dimension, referring to how action in relation to nature is always anchored in social structures and the conflicts that may emerge from these—such as, for instance, conflicts related to class or relations between the Global North and South, central or peripheral regions, or city and countryside. Nature interests do not emerge 'objectively' from structural conditions, but need to be articulated. While there is no pre-defined relation between a particular structure and the articulation of a nature interest, the former nevertheless *conditions* the latter. In the historical period that this book covers, we see a number of different articulations of particular class interests in the shape of nature interests. For example, the early twentieth-century conservation movement mobilized different sections of the bourgeoisie through the articulation of conflicting *instrumental* and *aesthetic* nature interests. While working-class interests can be mobilized in the shape of an instrumental nature interest, as when workers side with corporate interest to defend jobs in the fossil fuel industry, working-class interests can also be articulated through a *social* nature interest, as in the Green New Deal, where environmental concerns are linked to the creation of 'green jobs'. This means that the concept of nature interest can help us both to understand how internal movement tension is 'internalized' structural conflict and how such structural conflict may be negotiated and bridged by movement articulation. For example, in some cases a social and an ecological nature interest may be linked without friction in movement mobilization, while in other cases they may give rise to an internal struggle.

By discerning these three dimensions of nature interest, we will analyse how the environmental movement's conceptions of and emotions regarding nature interact with its strategies (including its forms of organizations) and the structural contexts in which it acts. The concept of nature interest thus becomes an important tool to understand how movement narratives articulate both continuity and change in the history of the environmental movement and how it relates to knowledge production, action strategies, and social structure as well as broader processes of structural change.

Critique of the Instrumental Nature Interest

The usefulness of distinguishing between a concept of nature and a nature interest is perhaps most clear when comparing the instrumental concept of nature with what we call an instrumental nature interest. While the former must include all humanity's uses of nature, from hunting, fishing, and small-scale farming to industrial capitalism's large-scale extraction of natural resources, *we use instrumental nature interest to refer to a systematic exploitation of nature that is ultimately subsumed to economic interest.* What unites the heterogeneous environmental movement historically as well as globally is its critique of the hegemonic instrumental nature interest. This emerged historically with industrial capitalism and found expression in *practices* in which nature *by definition* is treated as an instrument to reach economic goals, under which it is thereby subsumed. As pointed out by Marx in *Grundrisse* already in 1858, this means that

> [N]ature becomes *purely* an object for humankind, *purely* a matter of utility; ceases to be recognized as a power for itself; and the theoretical discovery of its autonomous laws appears merely as a ruse as to subjugate it under human needs, whether an object of consumption or as a means of production. (Marx, 1973, p. 410, *our italics*)

This instrumental nature interest is rejected by all environmental narratives, including that of green progress. The latter relies on an instrumental *concept of nature* but not on an instrumental *nature interest*. An instrumental concept of nature is inherent in the idea that nature may be used to provide resources for society, although to which degree this should be done may vary greatly in different narratives. The narrative of green progress stresses that a cautious exploitation of natural resources without destroying 'ecological balance' is possible. An instrumental concept of nature is thereby subordinated to an overarching goal of 'ecological balance' or 'sustainability'.

Varieties of Nature Interest

In their opposition to instrumental nature interest, various sections of the environmental movement have articulated different, and sometimes conflicting, nature interests that often coexist within the different currents of the environmental movement. The concept of nature interest helps us see how this movement develops through internal processes where these different interests sometimes support each other and sometimes generate tensions and debates in the movement. Here, we offer a brief overview of the nature interests that we address in the book:

The early conservation movement was animated by a concert of nature interests, all based in bourgeois or upper-class milieus and sensibilities. The articulation of a *scientific nature interest* was crucial, partly reflecting the prominent position of natural scientists in the early conservation movement. While science was indispensable for the industrial-capitalist breakthrough and the accompanying exploitation of nature, it was here turned into a critique of such exploitation through the argument that preserving nature was scientifically valuable.

The value of nature could also be defined in aesthetic terms. The *aesthetic nature interest* was prominent in early conservation movements, being formulated along two ideological lines: anti-commercialism and nationalism, both of which expressed fundamentally bourgeois sensibilities. Anti-commercialism was expressed in the frequent criticism of how modern infrastructure and mass tourism 'uglified' nature. At the same time, the beauty of nature was ideologically central to the aestheticization of politics by nationalist currents, which grew stronger in the first decades of the twentieth century.

A *recreational nature interest* was also central to nature conservation. The starting point for its emergence was the division between the spheres of work and leisure that was established with capitalist production. The recreational nature interest appealed to both working- and middle-class needs, ultimately aiming to recreate a broken body, or an exhausted soul, and thereby to restore the vitality necessary for productive labour.

Outside the bourgeois milieus typical of the nature conservation movement, we find a *moral nature interest*. It may be formulated in a secular version, as in the early twentieth-century life reform movement, which was anchored in the lower middle class, and a century later in new social movements, such as animal rights. Here nature is viewed as a source of moral development, which human beings should cultivate through self-discipline and an action-oriented 'natural' life. It may also be articulated in an ethico-religious variant as in indigenous struggles. Here nature is seen as the source of a divine moral order violated by industrial development, and defending this order becomes the basis for protests against companies and authorities that are violating the spiritual laws of nature.

The environmental movement of the post-war period arose in conjunction with ecological thinking and the realization that society and nature were inextricably linked in a larger planetary whole. In connection with this holistic understanding, two nature interests became prominent in the environmental movement. The *ecological nature interest*, in which nature is seen as a planetary system, originated in natural science, was embraced by sections of the new middle class, and is central to most of the well-known environmental organizations of the Global North. Here environmental action is ultimately aimed at restoring the ecological balance that has been seriously disturbed by environmental destruction. The *social nature interest*, which has post-war roots both in green branches of the labour movement and in the 1980s environmental justice movement, has gained new ground with the global climate justice movement in recent decades. It focuses attention on the devastating consequences that the abuse of nature has for the fabric of society itself. Since that abuse is seen as linked to the unequal distribution of opportunities to use nature and its resources, the fight against environmental degradation is also inextricably linked with a social justice perspective. In contrast to the biocentrism of the ecological perspective, the social interest in nature represents an anthropocentric perspective on the relationship between society and nature.

Moving to the twenty-first century, the emissions trading scheme created in connection with the Kyoto Protocol helped establish *an economic nature interest* in the environmental movement. Here nature is conceptualized as an economic system (as expressed, for example, in the concepts of 'ecosystem services') and as prescribing acts and forms of organization aimed at marketizing the relationship between society and nature. With a strong base in sections of the economic elites and the middle class oriented towards ecological modernization, this economic nature interest is sometimes hard to distinguish from the instrumental nature interest which is embedded in the dominant institutions of industrial capitalism. In response to this, *a juridical nature interest* has been articulated in the currents of justice activism. Originating in the Global South, this interest finds expression today in the campaign for the rights of nature and in initiatives such as the International Tribunal for the Rights of Nature, which we will discuss in Chap. 4.

Structure of the Book

The book's structure is straightforward. In the following three chapters, we devote one chapter to each of the three major environmental movement narratives—green progress, apocalypse, and postapocalypse. We present the central traits of the narrative (including its main ambivalences and contradictions) and the ideas and emotions that it expresses, to which it appeals, or which it attempts to evoke. Our historical approach involves an emphasis on those moments when the three narratives emerge as established features of environmentalism. In Chap. 2, 'Green Progress', we pay particular attention to the narrative of green progress in the early twentieth-century nature conservation movement; in Chap. 3, 'Apocalypse', we focus on how the apocalyptic narrative is established in the decades following the Second World War; and in Chap. 4, 'Postapocalypse', we highlight how the postapocalyptic narrative gains ground in the twenty-first-century climate movement. We also discuss how the narratives have developed within the environmental movement, situating them in relation to relevant shifts in capitalist society and the forms of environmental destruction that have constituted their main focus. We also exemplify how the narratives have varied in relation to class and social geography (e.g. Global North and Global South). Finally, each chapter also points to the major criticisms that have been levelled at each narrative and the major debates that they have occasioned. Empirically, these chapters rely on both secondary literature and empirical material—movement texts as well as interviews and observations—gathered in our own research projects about the environmental movement in the last decade (e.g. Cassegård et al., 2017; Cassegård & Thörn, 2018). This, however, is not primarily an empirical book, but one that sets out to make a historically informed theoretical and political argument. While we have made efforts to include perspectives of movements based in the Global South in our empirical research (to which, in addition to organizations based in Africa, Asia, and Latin America, we count indigenous movements based in countries like Japan and Sweden), we acknowledge a certain Global North bias in our material.[1] Nevertheless, the theoretical argument made here is defined by an intent to avoid such a bias. A fundamental ambivalence in the postapocalyptic narrative, the centrepiece of our book, revolves around what forms of action remain meaningful despite the disasters, and what space remains for such action. How can activism be sustained despite the loss of hope in the idea that catastrophe can be averted? What forms of political conflict become central when catastrophe is no longer just a future threat but a present reality? These are questions which we ourselves take up in the final chapter, 'Towards a Critique of the Environmental Movement'. Here we formulate a critical analysis of the environmental movement, returning to the question why it has failed to halt the environmental degradation that today defines life on earth. Inspired by the critical procedure of the early Frankfurt School, which combined ideology critique with a concern for utopia, we turn to a critical analysis of the limitations and possibilities of the three narratives.

References

Adorno, T. W. (1973). Die Idee der Naturgeschichte. In T. W. Adorno (Ed.), *Ges. Schriften Bd I, Philosophische Frühschriften* (pp. 345–365). Suhrkamp.

Bakhtin, M. (1981). *The Dialogic Imagination: Four Essays.* University of Texas Press.

Benjamin, W. (1985). *The Origin of German Tragic Drama.* Verso.

Benjamin, W. (1991). Das Kunstwerk im Zeitalter seiner technischen Reproduzierbarkeit (Erste Fassung). In W. Benjamin (Ed.), *Gesammelte Schriften 1.2* (herausg. v. R. Tiedemann u. Hermann Schweppenhäuser, pp. 435–469). Suhrkamp.

---

[1] In the research project Climate Action in a Globalizing World, we conducted 145 formal and 66 informal interviews at the COP meetings in Warsaw (2013), Lima (2014), Paris (2015), and in Japan, Denmark, Sweden, and the US. The empirical material of this project also included observations during the three COP meetings, on the UN Summit on sustainable development in Rio de Janeiro in 2012 as well as in connection with a number of other environmental movement meetings in the four mentioned countries (see Cassegård et al., 2017, p. 245ff.). We also make use of material from an ongoing research project, Adapting to Climate Change: Emotions and Narratives in the Environmental Movement, mainly 32 interviews conducted in 2020–2021. Cassegård and Thörn (2023) is a book on the history of the environmental movement in Japan and Sweden (1870–2020), mainly based on analyses of movement texts (including journals, manifestos, and books by movement intellectuals).

Benjamin, W. (1997). Some Motifs in Baudelaire. In *Charles Baudelaire: A Lyric Poet in the Era of High Capitalism* (pp. 107–154). Verso.

Berger, J. (1999). *After the End: Representations of Post-Apocalypse*. University of Minnesota Press.

Cassegård, C. (2021). *Toward a Critical Theory of Nature: Capital, Ecology, and Dialectics*. Bloomsbury.

Cassegård, C., Soneryd, L., Thörn, H., & Wettergren, Å. (Eds.). (2017). *Climate Action in a Globalizing World: Comparative Perspectives on Environmental Movements in the Global North*. Routledge.

Cassegård, C., & Thörn, H. (2018). Toward a Postapocalyptic Environmentalism? Responses to Loss and Visions of the Future in Climate Activism. *Environment and Planning E: Nature and Space, 1*(4), 561–578.

Cassegård, C., & Thörn, H. (2023). *I skuggan av apokalypsen: miljörörelser och industrikapitalism 1870–2020*. Göteborg: Daidalos.

Commoner, B. (1971). *The Closing Circle: Confronting the Environmental Crisis*. Jonathan Cape.

Czarniawska, B. (2004). *Narratives in Social Science Research*. Sage.

de Moor, J., & Wahlström, M. (2019). Narrating Political Opportunities: Explaining Strategic Adaptation in the Climate Movement. *Theory and Society, 48,* 419–451.

Eyerman, R., & Jamison, A. (1991). *Social Movements: A Cognitive Approach*. Polity Press.

Fanon, F. (1987). *White Skin, Black Masks*. Pluto Press.

Habermas, J. (1971). *Knowledge and Human Interests*. Beacon Press.

Hadden, J. (2017). Learning From Defeat: The Strategic Reorientation of the U.S. Climate Movement. In C. Cassegård, L. Soneryd, H. Thörn, & Å. Wettergren (Eds.), *Climate Action in a Globalizing World: Comparative Perspectives on Environmental Movements in the Global North* (pp. 143–164). Routledge.

Hamilton, C. (2017). *Defiant Earth: The Fate of Humans in the Anthropocene*. Polity Press.

Jacobs, J. (1992). *The Death and Life of Great American Cities*. Vintage Books.

Jameson, F. (1981). *The Political Unconscious: Narrative as a Socially Symbolic Act*. Cornell University Press.

Jamison, A. (2001). *The Making of Green Knowledge*. Cambridge U.P.

Kaplan, A. E. (2016). *Climate Trauma: Foreseeing the Future in Dystopian Film and Fiction*. Rutgers University Press.

Killingsworth, M. J., & Palmer, J. S. (1996). Millennial Ecology: The Apocalyptic Narrative from *Silent Spring* to *Global Warming*. In C. G. Herndl & S. C. Brown (Eds.), *Green Culture: Environmental Rhetoric in Contemporary America* (pp. 21–45). University of Wisconsin Press.

Klein, N. (2000). *No Logo*. Flamingo.

Laclau, E. (1990). *New Reflections on the Revolution of Our Time*. Verso.

Lukács, G. (1971). *History and Class Consciousness: Studies in Marxist Dialectics*. Merlin Press.

Malm, A. (2016). Who Lit This Fire? Approaching the History of the Fossil Economy. *Critical Historical Studies, 3*(2), 215–248.

Marx, K. (1973). *Grundrisse*. Penguin Books.

McAdam, D., Tarrow, S., & Tilly, C. (2001). *Dynamics of Contention*. Cambridge University Press.

McKibben, B. (1989). *The End of Nature*. Random House.

Melucci, A. (1989). *Nomads of the Present: Social Movements and Individual Needs in Contemporary Society*. Temple University Press.

Methmann, C. P. (2013). The Sky Is the Limit: Global Warming as Global Governmentality. *European Journal of International Relations, 19*(1), 69–91.

Moore, J. W. (Ed.). (2016). *Anthropocene or Capitalocene? Nature, History, and the Crisis of Capitalism*. PM Press.

Rancière, J. (1999). *Disagreement: Politics and Philosophy*. University of Minnesota Press.

Ricoeur, P. (1986). *Lectures on Ideology and Utopia*. Columbia University Press.

Smith, N. (2010). *Uneven Development: Nature, Capital and the Production of Space*. Verso.

Soper, K. (1995). *What is Nature? Culture, Politics and the non-Human*. Blackwell.

Stoner, A. M., & Melathopoulos, A. (2015). *Freedom in the Anthropocene: Twentieth-century Helplessness in the Face of Climate Change*. Palgrave Macmillan.

Swyngedouw, E. (2013). Apocalypse Now! Fear and Doomsday Pleasures. *Capitalism Nature Socialism, 24*(1), 9–18.

Thörn, H. (1997a). *Rörelser i det moderna. Politik, modernitet och kollektiv identitet i Europa* (pp. 1789–1989). Rabénförlagen.

Thörn, H. (1997b). *Modernitet, sociologi och sociala rörelser*. Department of Sociology, University of Gothenburg.

Thörn, H. (2015). How to Study Power and Collective Agency: Social Movements and the Politics of International Development Aid. In S. Hansson, S. Hellberg, & M. Stern (Eds.), *Studying the Agency of Being Governed: Methodological perspectives* (pp. 85–102). Routledge.

Thörn, H. (2019). Revolution as a Politics of Time-Space: From Enlightenment Modernity to Advanced Globality. In E. Namli (Ed.), *Future(s) of the Revolution and the Reformation* (pp. 65–95). Palgrave Macmillan.

Thörn, H. (2022). New Forms of Environmental Movement Institutionalization: Marketization and the Politics of Responsibility. In I. M. Grasso & M. Giugni (Eds.), *The Routledge Handbook of Environmental movements* (pp. 325–337). Routledge.

Thörn, H., & Svenberg, S. (2016). 'We Feel the Responsibility That You Shirk': Movement Institutionalization, the Politics of Responsibility and the Case of the Swedish Environmental Movement. *Social Movement Studies, 15*(6), 594–609.

Thunberg, G. (2019). *No One is Too Small to Make a Difference.* Penguin.

Wapner, P. (2010). *Living Through the End of Nature: the Future of American Environmentalism.* MIT Press.

Warde, P., Robin, L., & Sörlin, S. (Eds.). (2018). *The Environment: A History of the Idea.* Johns Hopkins University Press.

Williams, E. C. (2011). *Combined and Uneven Apocalypse.* Zero Books.

CHAPTER 2

Green Progress

Abstract This chapter introduces the green progress narrative, the core of which is belief in the compatibility of environmental protection and economic growth. The narrative can be traced back to the early twentieth century nature conservation movement. Today it is commonly articulated in connection with the notions of sustainability and 'green' growth. While predominantly aligned with neoliberal globalism, a socialist version of the green progress narrative has gained salience in recent years through the Green New Deal. A recurring ambivalence in the narrative stems from its attempt to present human progress as historically necessary while at the same time stressing the need for action to ensure further progress in the future. Emotionally, this ambivalence is expressed as a tension between certitude and hope.

Keywords Environmental movement • Narrative • Progress • Growth • Green new deal • Conservation movement

BETWEEN CERTAINTY AND HOPE

The narrative of green progress is emotionally centred around an ambivalence between certainty and hope. Hope is in this context an anticipation of a better and more developed society, motivating future-oriented action, which is linked to a belief that history is imbued with an evolutionary force

© The Author(s), under exclusive license to Springer Nature Switzerland AG 2022
C. Cassegård, H. Thörn, *Post-Apocalyptic Environmentalism*,
https://doi.org/10.1007/978-3-031-13203-2_2

that manifests itself in economic and technological development. Conversely, certainty about progress as a historical necessity may also function to sustain hope in the meaningfulness of action in times of movement defeat. Born in the early twentieth-century conservation movement, the narrative of green progress is a variation of—and a correction to—the Enlightenment progress narrative, which in the nineteenth century legitimated the breakthrough of industrial capitalism through a celebration of scientific development, technological expansion, and the economic growth generated by capital accumulation. The green progress narrative emphasizes that these forces need to be controlled or else they risk destroying nature. The insight that human progress comes with risk and therefore needs to be steered means that the narrative of green progress involves a stronger emphasis on agency than the Enlightenment version. However, the narrative of green progress did not question technological and scientific development, and certainly not the power relationships that were established with the emerging new social order of capital. On the contrary, it was anchored in sections of the new bourgeoisie who were as much alarmed by environmental destruction as by the anti-modernists who embraced nature conservation from a radically different angle. This meant that the green progress narrative was born out of a sometimes bitter internal struggle in the early conservation movement between *preservationists* and *conservationists*. Based mainly on an aesthetic nature interest, involving a romantic concept of nature, the preservationists argued that the ultimate aim of the movement was to ensure that nature was protected from further human intrusion. The conservationists, on the other hand, defended the idea that modern society's relation to nature needed to be regulated and managed in order to secure the possibility of extracting natural resources without destroying nature. Its utilitarian approach was also manifested in the fact that it often promoted a recreational nature interest. While conservation thus implied an instrumental concept of nature, it rejected the practices of an instrumental nature *interest* for which nature was nothing more than a pool of potential resources that—without any restrictions—could and should be exploited for the benefit of economic profit and/or growth. The conservationists saw limits to such exploitation of nature.

Parallel to the conservation movement, a number of other groups and organizations which should be understood as part of the early environmental movement sprang up in countries such as the US (Fox, 1981), Europe (Krabbe, 1974; Sharma, 2012; Treitel, 2017), and Japan (Hisai,

2018) in the first decades of the twentieth century, including the back-to-the-land and country life movements, the vegetarian movement, the nudist movement, the animal rights movement, the Georgist land reform movement, and the garden city movement. German historian Wolfgang Krabbe (1974) has coined the term 'life reform movement' as an umbrella for this heterogeneous wing of environmentalism. Similar to the conservation movement, the life reform movement involved a struggle between anti-modernists and modernists. In the latter case, an alternative version of the green progress narrative, emphasizing small-scale development, was formulated.

The narrative of green progress began to lose its foothold in the environmental movement after the Second World War. As we will see in Chap. 3, at this time an apocalyptic narrative was articulated by some of the leading intellectuals of the conservation movement. This development was fuelled by the emergence of the new environmental movement and its ecological nature interest, which meant that by the 1970s, the narrative of green progress was seriously weakened in the environmental movement. The narrative was then revived in the 1990s, as the movement embraced the discourse of sustainable development, followed by support for market solutions to address climate change in the wake of the Kyoto Protocol. Accordingly, this chapter will focus on the emergence of the green progress narrative in the early twentieth century and its revival a century later, both in a dominant neoliberal variant and in socialist versions.

Nature Conservation in the Early Twentieth Century: The Birth of the Narrative of Green Progress

The historian Richard Grove (1995) has criticized the idea that environmental thinking and ideas of nature conservation began in Europe and North America. Instead, these ideas were born in the colonies, when botanists and other natural scientists observed how colonialism devastated vulnerable ecosystems and learnt from indigenous knowledge about how to preserve nature while also using it to secure a livelihood. These experiences were brought back to the Global North and provided important impulses to the formation of the organized conservation movement, which was first established in the US in the 1890s, before spreading to Europe and East Asia in the early twentieth century.[1]

The struggle between preservationists and conservationists within the emerging nature conservation movement, from which the narrative of green progress emerged, was largely a battle over an aesthetic nature interest and the romantic concept of nature. In countries like the US and Sweden (but not in Japan), this also involved understanding, and romanticizing, nature as 'wilderness' (e.g. Wapner, 2010). This conceptualization of nature as the Other in Western culture has three historical roots: the Puritan 'frontier spirit' that was born in connection to the seventeenth-century colonization of North America; eighteenth-century bourgeois romanticism in Europe; and late nineteenth- and early twentieth-century Darwinism. From the view of the Puritans, wilderness was not associated with beauty but appeared as frightening and disharmonious, an alien force, resistant to human expansion (Merchant, 2004). It could only be subdued through dedicated and disciplined efforts in line with the Protestant ethic. While nature also represented the Other to civilization in the romantic movement, the latter charged the wilderness with aesthetic values. Wilderness not only represented beauty but was a source for experiencing 'the sublime' in its Kantian sense, that is, the experience through which we are confronted with an object that reveals the limits to what human sensibility and imagination can perceive and construct knowledge about. Moving to the moment of the birth of nature conservation, romanticism had been infused with Social Darwinism and the idea of natural history. This was clearly visible in the new literary genre that popularized vulgar Darwinism's survival of the fittest through narratives of small groups of white men who challenged and ultimately conquered the forces of nature. This was a discourse that not only celebrated masculinity, but also constructed racist perceptions about indigenous people, who were considered as part of the 'wild'. Nature conservation was deeply shaped by this discourse and its inner tensions, which often translated into conflict within the movement. Preservationists and conservationists alike challenged the translation of the conquering spirit (whether in its Protestant or Darwinist

---

[1] This chapter is first and foremost based on our own research on the early nature conservation movement in Sweden and Japan (Cassegård & Thörn, 2023), and on Stephen Fox's comprehensive history of the movement in the US. While the Swedish movement was highly influenced by the US movement, the movement in Japan displays characteristics that reflect a longer history of government-led forest conservation (preceded by a high degree of exploitation).

incarnation) to an instrumental nature interest, but nevertheless repro-
duced the racist elements of Social Darwinism in the shape of a paternalis-
tic exotism; indigenous people needed to be protected because they were
part of, or at least stood close to, nature. Further, in countries such as the
US and Sweden, nature conservation's perception of the 'untouched' wilderness that needed to be protected was also in some cases the result of a
misguided colonial gaze, as the celebrated areas of wild nature in fact had
been shaped by centuries of interaction with indigenous people (Fox,
1981; Cassegård & Thörn, 2023).

When the aesthetic nature interest emerged as a key element of early
nature conservation, it thus involved a romantic concept of nature,
anchored in a cultural sensibility associated with conservative sections of
the bourgeoise, and advocating the protection of nature from the destructive colonizing forces of industrial capitalism. In the debate between preservationists and conservationists, the aesthetic nature interest was
sometimes questioned by the latter. For example, Gifford Pinchot, the
leading voice of US conservationism, who eventually also promoted it
from the position of government under the Roosevelt administration, had
a deep dislike for the use of the word 'aesthetic' in connection with conservation (Fox, 1981, p. 115). But rather than denying that nature had
aesthetic values, conservationists tended to argue that preservationists
clung to it in a dogmatic manner, failing to realize that nature had other
values that needed to be considered. For example, an article in the Sierra
Club Bulletin, published in 1902, argued that 'there are aesthetic uses as
well as commercial uses—uses for the spiritual wealth of all, as well as for
the material wealth of some' (quoted from ibid., p. 115). It was precisely
the will to compromise that distinguished the conservationists from the
preservationists. Such pragmatism easily harmonized with the instrumental concept of nature that guided the former.

The Aesthetic and/or Recreational Values of Nature

When conservationists promoted an aesthetic nature interest, it involved
the action strategies of anti-commercialism and nationalism. The former,
most often embraced by preservationists, emphasized the need to fight the
'uglification' of nature that resulted from the commercial colonization of
nature. The latter, embraced by preservationists and conservationists alike,
used nature and natural values as symbolic resources in the construction of
national identity. In countries such as Japan, the US, and Sweden,

celebrating the nation's original nature and its specific beauty was an ideological cornerstone in early twentieth-century nationalism and its aesthetization of politics. In these countries, nature often appeared as the soul of the nation. In the context of nature conservation strategy and practice, this was manifested when demands for nature reserves were articulated through the creation of 'national parks'. For example, as head of the government US Forest Service, Pinchot renamed the 'forest reserves' as 'national forests'. In the US context, the frontier, understood as a primeval wilderness, could also be seen as the original source of American democracy and values (ibid., pp. 130; 116). However, even more important for utilitarian conservationism was the instrumentalist mode of defining nature as belonging to the nation in terms of 'resource nationalism'.

The beauty of nature as a national resource could also be linked to a recreational nature interest. The idea of recreation builds on the distinction between work and leisure time established in nineteenth-century bourgeois society. Leisure time is subordinated to labour time in the sense that the former is seen as a resource for recreating the energies needed for productive work. Nature is seen as a fountainhead of such energies; physical contact with nature could in a short space of time (over a weekend) recharge body and soul, rejuvenating the individual and preparing them to re-enter the workplace. Particularly when illness has struck body or soul, the direct experience of mountains, rivers or forests is seen as a key that can unlock the door to the pool of energies that reside in humans' inner nature. While this recreative nature interest began as a cultural expression and a privilege for bourgeois families, it became popularized in early twentieth-century nationalism, in which (national) nature was seen as a resource belonging to the people. In this context, it was linked to the birth of mass tourism, which began to involve the working class at the time of the rise of welfare states in the Global North. In the context of the Global South, consumption of nature based on a recreative nature interest was of course always a colonial phenomenon (especially in post-colonial times) and was thus never popular in any other sense than as an opportunity for income.

For the past century, the recreative nature interest was always a strong concern for conservationists, and thus a fundamental part of the green progress narrative. Initially preservationists were not hostile to the idea of nature as recreation; in the sense of a source for an inner, sublime experience, they did see a connection between the aesthetic and recreational values of nature. In Sweden, the major organizations for nature conservation and tourism joined forces (even establishing a shared board) in the 1910s (Haraldsson, 1987). Yet, this was not without tension. As tourism began to translate into *mass* tourism, a cleavage between the aesthetic and the recreative nature interests began to emerge, as preservationists protested the commercialization and uglification of nature brought by this development. As always, conservationists' pragmatic sense involved an attempt to strike a balance between different interests, always with an eye on utility in its multiple sense. For example, a programmatic piece in the first issue of the *Wilderness Society* (1935) defined nature as 'the environment of solitude' and as 'a human need rather than a luxury or a plaything', while providing a cautious defence of tourism based on a recreative nature 'excluding elaborate hotels, admitting responsible packers' (quoted from Fox, 1981, pp. 211; 214). Such pragmatism was even more pronounced when the Sierra Club sent a statement to a government conference on natural resources (organized by Pinchot): 'The moral and physical welfare of a nation is not dependent alone upon bread and water. Comprehending these primary necessities is the deeper need for recreation and that which satisfies also the esthetic sense' (quoted from ibid., p. 130).

Nature as a Resource for Electricity

In the US as well as in Sweden, decisive battles between preservationists and conservationists were played out in conflicts over hydropower. These conflicts were crucial for establishing the narrative of green progress in these contexts. This needs to be understood in relation to the fact that these battles occurred at a point in the early twentieth century when the fundamental dependence of industrial expansion on an increasing access to electricity became apparent to industrialists as well as politicians, including V. I. Lenin (2012, p. 419), who in 1920 argued that 'communism is soviet power plus the electrification of the whole country'. Thus, it was crucial for any nation in the process of industrialization to find new sources of electricity at as low a cost as possible. It was significant how Pinchot, in his defence of turning the valley of Hetch Hetchy (baptized by the indigenous people with reference to its grassy meadows) into a water reservoir, underscored both how the value of hydropower could not be underestimated and that it needed to stay in the hands of government (to avoid monopolistic control). Pinchot argued that '[w]hoever dominates power dominates all industry' (quoted in Fox, 1981, p. 141). To movement intellectuals and leaders like Pinchot, the ultimate goal of conservation

was to achieve the most efficient use of resources while at the same mini-
mizing the damage to nature. In both the US and Sweden, such 'damage
control' also involved avoiding causing any disturbance to the lives and
livelihoods of indigenous peoples as much as possible. Such efficiency was
linked to the concept of 'multiple use', which in the case of damming
involved the provision of drinking water, regulation of floods and erosion,
the irrigation of crops, and, most important, the production of energy. It
is even fair to say that, considering how the battle over hydropower was
followed by nature conservation's promotion of nuclear power in the
1950s and solar and wind energy in the 2000s, advocating 'clean' sources
of energy as an engine of societal development has been the most basic
storyline of the green progress narrative.

In such conflicts, a struggle over how to define a scientific nature interest became prominent. This points to how conflicts over nature also cut
through the scientific community; while scientists and engineers acted as
midwives of the instrumental nature interest, they were opposed by a few
colleagues who joined the nature conservation movement. The latter
articulated a scientific nature interest, arguing that nature needed to be
protected because it was *an invaluable source of knowledge*. An important
background to this claim is the idea that nature in the wake of Darwin's
discoveries is not a static phenomenon; it is constantly changing and thus
has a history. Studying nature is therefore always about studying its history, which is natural history that ultimately is also human history.
Destroying nature would thus erase the sources for studying and understanding the history of the planet and humanity's place in it. Given this
view, which was based on an evolutionary perspective, the scientific nature
interest could be mobilized by both preservationists and conservationists.
In the case of the former, scientists could argue that total protection was
required, sometimes in order to preserve remnants of previous stages in
nature's history. This approach involved the perspective of racist biology;
the cultures of indigenous peoples were worth preserving because they
represented historical stages in the evolution of humanity. For example, in
1913 *Sveriges Natur* (p. 1), the journal of the Swedish Conservation
Society, defined conservation as a demand that 'man protects nature (the
plants, the animals and the primitive peoples) against—man' (the literal
translation to Swedish of 'primitive people' is *naturfolk*—people of
nature).

When conservationists embraced a scientific nature interest, it most
often served to legitimize 'cautious exploitation'. Scientists were

mobilized to make sure that the exploitation of resources was exercised in a manner that took the scientifically defined values of nature into consideration. This often involved a different evolutionary (but no less racist) approach to indigenous peoples, as it was argued that they should be subjected to 'development efforts'. In practice, this could mean that when villages were flooded by damming in connection to the construction of hydropower, relocation involved placing indigenous children in schools and thus 'liberating' them from the bonds of traditional lifestyles (e.g. Cassegård & Thörn, 2023).

The perspective of natural history also opened the way for links between nature and culture conservation. For example, in the context of Swedish conservation, '*naturminnen*' and '*kulturminnen*' were framed in a similar mode, referring to small spaces of conservation that were designed to preserve ancient stages of natural and/or cultural history. Such spaces could thus appear both as laboratories for science and open-air museums for the public. This construction implied the concept of second nature, in which the boundaries between natural and human history are more or less dissolved. Such dissolution defined the Japanese nature conservation movement, in which the idea of nature as wilderness was absent. This needs to be understood in relation to a number of factors, including population density, a pre-modern history of forestry governed by the state and not least upper-class culture, in which forests were not associated with hunting but with ancestral spirits and the history of the imperial house. This also meant that in the Japanese context, the scientific nature interest was primarily linked to the humanities rather than natural science (Cassegård & Thörn, 2023).

From Conservation to Nature Engineering

In his analysis of the controversy around Hetch Hetchy within the conservation movement, Stephen Fox (1981) highlights how those pushing for the dam were politicians and professionals employed by San Fransisco City. In connection to this, he argues that the conflict between preservationists and conservationists in the early twentieth century was not just a battle of ideas and strategies, but also needs to be understood in relation to social structure. The preservationists were driven by (people who could afford to be) amateurs, while the conservationist branch was led by professionals, often working in local, regional or national government. In a country like Sweden, the preservationist approach, at least in the early phase of the conservation movement, had strong links to the landed gentry and its hunting culture, while bourgeois professionals tended to dominate among the conservationists.

This predominance of professionals amongst conservationists also explains how their victory over the preservationists was the start of a process through which conservation was more or less turned into a dimension of state ideology in the inter- and post-war periods. In the US, this process meant that while the conservation movement started out as predominantly conservative, it ended up as a social-liberal project in the inter- and post-war period. In fact, considering the significance of a conservation perspective in the political project led by Franklin D. Roosevelt, it is fair to say that there was a strong green element already in the 1930s New Deal (in the 1930s, the conservation issue shifted from being a concern of the Republican Party and business interests to that of the Democrats and government, ibid., pp. 185–186). The process of translating the green progress narrative into state ideology did not only occur in the US; it also happened in other countries where the state introduced efforts to embed capitalism (Polanyi, 2001). In a sense, this meant that state-managed capitalism in the Global North (including Japan) was linked to, and even presupposed, nature engineering. Thus, in the second half of the twentieth century, the hegemonic status of the green progress narrative meant that the values of amateurism and protectionism had to give way to professionalism, bureaucratization, scientific management, planning, efficiency, utilitarianism, and nature as a commodity.

Nature as a Source of Moral Development: The Life Reform Movement

The life reform movement (Krabbe, 1974, 1998; Sharma, 2012; Treitel, 2017) did not celebrate the wilderness, but rather the countryside, with its long history of human cultivation of nature. More clearly than nature conservation, the life reform movement articulated an experience of unhappiness about life in the modern city; while the wilderness first and foremost was a place that civilized man could visit and worship, the countryside represented a place to live an everyday life in harmony with nature. The life reform movement shared with nature conservation a certain nostalgic conception of nature, with nationalist connotations; the countryside refers to *home* (more clearly pronounced in the German *Heimat* or the

Scandinavian *hembygd*), which is a place for the individual to return (back home) to, and which at the same time represents the nation. Similar movements arose in Japan where they were centred on notions such as *furusato* ('old village'/'native region'). By returning to the country, the individual would be reunited not only with her true inner self but also with the soul of the nation.

Radically different from the romantic concept of nature as wilderness, the concept of the countryside does not involve a sharp boundary between nature and society. On the contrary, nature and society do not appear as separate but as interwoven, a result of millennia of human cultivation of nature. This means that nature is perceived as 'co-produced' through interaction with society, and that nature has co-produced society, shaping it to such a degree that historically constructed cultural and social institutions appear as an independent external or objective reality: a second nature.

In the life reform movement, the concept of second nature was articulated as part of a moral nature interest, as nature was perceived as the source and guide of the 'natural lifestyle', which would steer humanity back into a harmonious relation with nature that had been lost with the advent of industrial capitalism. This means that the main form of action practised by this movement can be understood as what we call a life form politics (Thörn, 1999). In life form politics, the countryside represents an organic relation between nature and society. Moral nature interest is here opposed to the nature interests of nature conservation; 'wilderness' is not an ideal, it is not what is 'natural'. Natural life is nature domesticated, moulded by a humanity that understands how to cultivate without destroying. The bright open landscape, rather than the dark forest, was the space of human freedom for the life reform movement. Here, humanity can realize its true self through hard work in interaction with nature. The natural way of life appears as a life in second nature because everyday life is fully integrated with (first) nature. This also signals harmony between inner and outer nature. The natural way of life, however, does not arise spontaneously for those moving back from the city to the countryside. Modern man has to re-learn how to live naturally and must work hard to achieve this. Thus, the overarching goal of the life reform movement is to guide modern man back to the natural way of life.

According to the moral nature interest of the life reform movement, the individual must subordinate herself to self-discipline. Similar to the practice of agriculture, inner nature requires domestication. In a sense, this involves returning to nature but not as a 'savage' engaged in a struggle for survival, but as a morally refined individual who reveres, nurses, and nurtures nature. Nature has an inherent moral order, yet it is one that needs to be discovered and refined. Nature should thus not be subordinated to humanity; on the contrary, humanity should be guided by nature as a source of moral development. The utopia of the natural way of life constructs an image of an acting subject that enters into a deep relation with nature that at the same time involves moral education, since nature harbours seeds for a universal moral, which also provides a foundation for the development of human civilization towards a goal of planetary harmony.

This discourse frequently referred to health both literally and metaphorically. Returning to the nature of the countryside was a process through which humanity could be *healed* of the disease of modern life; when the unity of society and nature had been torn apart, individuals and societies had also been internally sundered. In concrete terms, the movement produced leaflets and books providing detailed instructions on how to live the natural way of life, which would lead to improved health both physically and psychologically. Such instructions importantly included pursuing a 'natural diet' and abstaining from meat, alcohol, tobacco, vaccines, and other 'unnatural' chemical cures. This was perhaps most clearly articulated in the vegetarian movement, which in Europe had an important base in Germany (Sharma, 2012; Treitel, 2017). Krabbe (1974) argues that vegetarianism was the core of the life reform movement. Here, eating meat was regarded as a civilizational delusion; human beings, according to their biological nature, are animals that should live on plants. The life reform movement also involved collective nowtopian action forms (Thörn, 1997, 2019), i.e. practising utopia here and now, as small groups of people left cities to found communities on the countryside dedicated to the natural way of life.

The life reform movement shared with other contemporary movements a critique of modernity that was directed at how both industrial technology and, first and foremost, urban life had put a distance between humanity and nature. In some cases, the movement was anti-capitalist but more often it was against (Western) civilization. As already mentioned, there were, however, tensions between a more anti-modernist wing (which overlaps with preservationism as celebrations of the countryside were to be found in journals of the conservation movement) and a modernist variant that was allied with the peace, women's, socialist, and animal rights movements of the early twentieth century. Elements of the former fed into the Nazi movement in the 1920s. This dark current of the movement's legacy is perhaps one reason why the life reform movement is seldom mentioned as part of the history of environmentalism. The left wing of the movement, however, was clearly a predecessor to the counterculture of the new social movements that began to emerge in the 1960s, as well as the more technologically oriented eco-villages of the 1990s. In a similar mode, the future-oriented section of the life reform movement articulated an early version of this green progress narrative, emphasizing small-scale development and decentralization along the lines of Pyotr Kropotkin's books *Fields, Factories and Workshops* (1898) and *Mutual Aid: A Factor of Evolution* (1902), which had a significant influence on the movement.

Through articulating a variant of the green progress narrative emphasizing human development as *organic* growth, the progressive sections of the life reform movement distanced themselves both from growth understood primarily in the economic sense, and from the conservative pastoral ideals of the static *heimat*. The lead character of this narrative is humanity as a farmer or a gardener, and the action forms prescribed are all defined as different forms of healing through growing. This could involve small-scale technology as long as it was subsumed under the goal of creating a decentralized society in harmony with nature.

This version of the green progress narrative was anchored in mobile and ambivalent class positions. Research on the vegetarian movement demonstrates how it was an urban phenomenon. In Germany, its strongholds were cities with a population that exceeded 100,000 (Krabbe, 1974; Stolare, 2003). The emergence of the movement in different countries tends to coincide with the breakthrough of industrial capitalism. The life reform movement's version of the green progress narrative thus first and foremost spoke to an experience of the mobility of urbanization. The nostalgia inherent in the life reform movement's longing for the countryside resonated with great numbers of people in the Global North in the late nineteenth and early twentieth century who lived in a city but had memories, and experienced the loss, of a life with clean air, a closeness to forests and lakes and rivers with clear water. Still, such longing seems chiefly an expression of the sensibilities of sections of the middle class. In Sweden, there was a strong difference between the vegetarian movement and the nature conservation movement in terms of social anchoring; while the latter was dominated by the upper middle class, the former was mainly anchored in the lower middle class, with a strong over-representation of elementary school teachers (Stolare, 2003, p. 101). The life reform movement addressed both peasants and workers, albeit without great success, with the exception of an overlap with the early twentieth-century youth movement, which attracted working-class youth (ibid., p. 2003). In a sense, the different social dimensions of nature conservation and life reform could perhaps be interpreted as partly reflecting varying experiences of urban life; while the successful sections of the bourgeoise were happy to return to their lucrative professions in the city after a weekend or a week 'rusticating' in 'wild nature', the life reform movement's longing to return to a permanent life in the countryside addressed those who did not fare quite so well in the city and who did not have the means to either extricate themselves from the more unhealthy urban environments or leave the city for longer or shorter holidays.

NATURE AS AN ECONOMIC SYSTEM: SUSTAINABLE DEVELOPMENT AND THE GLOBAL CONSOLIDATION OF NEOLIBERALISM

During the two first decades following the Second World War, the green progress narrative continued to occupy a dominant position in the environmental movement. While the conservation movement changed its position on hydropower in this period (Fox, 1981; Cassegård & Thörn, 2023), this came with the support for nuclear power as an alternative 'clean' source of energy. The process of institutionalization of the conservation movement and its narrative of green progress, however, gave birth to a slowly emerging counter-reaction within the movement in the postwar period. This involved the articulation of an apocalyptic narrative by some of its leading intellectuals (see Chap. 3), and in the 1960s it was also fuelled by the emergence of the new environmental movement and its ecological nature interest. For example, in both the US and Sweden, this internal 'radicalization' in the 1970s and 1980s saw the movement changing its previous affirmative positions on nuclear energy. In Japan, conservationism was challenged in the 1960s by growing public concerns about pollution and a wave of anti-pollution protests, which necessitated a shift of attention to social issues and social justice. This meant that the narrative of green progress was seriously weakened in the environmental movement. It was only in the 1990s, as the movement embraced the discourse of sustainable development, followed by support for market solutions in response to climate change in the wake of the Kyoto Protocol, that the green progress narrative re-emerged and again became highly influential within the environmental movement. This went hand in hand with the global establishment of sustainability discourse, advances in environmentally-friendly technologies (such as solar energy and electric cars), and increasing stress on market mechanisms and ecosystem services in environmental policy—with the neoliberalization of the global economy serving as an important structural prerequisite.

The 1992 Earth Summit (the UN Conference on Environment and Development) in Rio de Janeiro was a watershed in the comeback of the green progress narrative. An important background to the conference was the Commission on Environment and Development appointed by the UN General Assembly in 1983. Led by the Norwegian Social Democrat Gro Harlem Brundtland, the Commission launched the concept of *sustainable development* in its 1987 report *Our Common Future*—a concept that in the decades following the conference became linked to an astonishing range of policy programmes worldwide. The conference saw the birth of the UNFCCC (United Nations Framework Convention on Climate Change) as well as the Convention on Biological Diversity. The UNFCCC in turn led to the annual climate summits known as COP (Conference of the Parties), which produced the 1997 Kyoto Protocol and the 2015 Paris Agreement. Another fruit of Rio was Agenda 21, a comprehensive programme for changing consumption and production patterns, especially in the richest parts of the world.

Environmental activists had reason to regard the Rio conference with optimism. The actions of the past decades to protect the ozone layer had inspired hope that the Rio Declaration would be backed by global political action. It was in the early 1970s that scientists had discovered that the use of chemical substances, most notably freons, threatened the ozone layer of the atmosphere. The shocking discovery of a large ozone hole over Antarctica in the mid-1980s contributed to the apocalyptic fears of that decade, but the political response was relatively swift. The so-called Montreal Protocol (Montreal Protocol on Substances that Deplete the Ozone Layer), created in 1987, is often cited as one of the most successful international agreements ever written.

At the same time, a clear dividing line existed between large parts of the environmental movement and the political establishment around the question of whether economic growth should continue to be a beacon of global social development. The Brundtland Commission did not hesitate on this matter: it even demanded that the 'international economy must speed up world growth' (WCED, 1987, p. 89). The commission's approach was in line with the refrain of the green progress narrative about the possibility of harmonizing nature conservation and economic development. Importantly, it also coincided with the worldwide swing towards a full-scale neoliberal policy, which quickly incorporated the sustainability concept. With the fall of the Berlin Wall and China's embrace of a capitalist market economy, a political climate came into being in which think tanks and editorials could ridicule anti-capitalist ideas as obsolete, including the idea that capitalism was environmentally destructive. The new mantra was instead that societies should grow themselves out of the ecological crisis and that problems such as global warming should be solved by innovation and entrepreneurship in combination with market mechanisms such as emissions trading. Legitimizing this mantra was the so-called 'ecological modernization' approach (e.g., Mol & Spaargaren, 2000), which suggested that environmental problems rooted in previous forms of modernization could be solved by 'ecological enlightenment' (Beck, 1995).

This optimistic belief in markets and technology forms the core idea of the contemporary green progress narrative, which from the 1990s onwards has gained a firm foothold in those established environmental organizations that Naomi Klein (2014) refers to as the 'Big Green' and that includes well-known names such as Greenpeace and the World Wildlife Fund. With the support of these organizations for the emissions trading established through the Kyoto Protocol, an *economic nature interest* was now introduced into the environmental movement's green progress narrative for the first time, where it coexisted with and often subsumed the ecological and social nature interest.

Neoliberal globalization furthered the established environmental organization's embrace of the green progress narrative in at least two ways. First, the relocation of industrial production and other environmentally destructive economic activities to the Global South was accelerated. The shift towards a seemingly post-industrial economy in the Global North made it easier for these countries to present themselves as pioneering green countries despite their continued dependence on resource extraction and polluting production globally. Second, globalization was accompanied by the emergence of a global climate regime that took shape after the Rio conference. The new global governance arrangements included intensified cooperation between authorities and environmental organizations, as well as increased reliance on market mechanisms such as emissions trading. Under an overarching framework defined by nature's economization, the state shifted part of the responsibility for solving environmental problems to consumers, businesses, and environmental organizations. With this, the institutionalization of the environmental movement both accelerated and changed shape, in the sense that contacts and collaborations with not just authorities but now also corporations became a regular part of the activity of many established environmental organizations (Thörn & Svenberg, 2016).

Nature and Social Justice: The Green New Deal

The last decade has also seen the establishment of a new socialist-oriented version of the green progress narrative, best exemplified by the Green New Deal but also evident in the technological utopianism of works such as Aaron Bastani's *Fully Automatic Luxury Communism*, published in 2019, and the 'accelerationist' Left with its calls for a 'Promethean politics of maximal mastery over society and its environment' (Williams & Srnicek, 2013). Here, as in the neoliberal variant of the narrative, there is a strong belief in the possibility of sustained growth and in the vision of a high-tech world with a vast majority of the world's population living in 'green' big cities. In socialist versions of the green progress narrative, however, the economic and ecological nature interests are subordinated to an overriding social nature interest, as signalled by the emphasis on 'green jobs' and the slogan of a 'just transition', strongly rooted in the climate-conscious parts of the international trade union movement.

The Green New Deal is best known in the variant pushed for by Alexandria Ocasio-Cortez and Ed Markey in the US Congress in 2019 with the strong backing of the Sunrise Movement. It combines drastic measures against the fossil fuel industry with massive investments in green technology, anti-poverty measures, and the creation of green jobs, all done to steer clear of climate catastrophe while furthering social justice. Naomi Klein called it 'a once-in-a-century chance' to fix the failed economic model, reserving specific praise for the social components of the reform programme—creating 'hundreds of millions of good jobs', investing in systematically excluded communities and nations, and guaranteeing healthcare and childcare (Klein, 2020, p. 26).

Next to the prominent place of the social nature interest, the stress on the need for mass mobilization is the most distinguishing mark of this Leftist variant of the green progress narrative. This is especially evident in versions of the Green New Deal embraced by radical movement intellectuals, such as Alyssa Battistoni. In the 2019 book *A Planet to Win: Why We Need a Green New Deal*, she and her co-authors argue that a quick decarbonization of the economy will only be possible through massive mobilizations, including by the labour movement. Here a clear awareness exists that the transition will inevitably lead to resistance from the fossil fuel industry and other elites, and that strong movement support is indispensable. This in turn can only be achieved by showing how solving the climate issue offers the chance to dismantle inequalities and improve the lives of common working people (Aronoff et al., 2019). Klein concurs, arguing that the social and economic aspects of the Green New Deal, far from weighing the proposal down and making it more impractical, are in fact 'precisely what is lifting it up' by making it attractive to the working class (Klein, 2020, p. 288).

By emphasizing conflict and mobilization, this variant of the green progress narrative downplays the idea of progress as a historical necessity. Instead, it is seen as a possibility in which the masses are given a key role as active agents. Here we recognize the central ambivalence of the green progress narrative which we pointed to in Chap. 1, namely its tendency to oscillate between a view of progress as inevitable and as requiring (sometimes heroic) agency. A similar ambivalence can be found in the manifestos of Bastani and Left accelerationism (Gordon, 2022). That this is a progress narrative is nevertheless shown by the fact that it is through a promise of improvements towards a bright future that it mobilizes supporters, rather than through doomsday threats. Much of *A Planet to Win* is thus taken up with presenting an attractive vision of the future which is not only greener but one in which society has also become more just and the masses enjoy material prosperity and even 'abundance' and 'universal luxury'—the latter a result of redistribution policies, job guarantees, division of labour, and 'smart' technology. Everyone will thus live in 'lovely' public housing, with 'smart' appliances and 'smart' grids, with access to speedy trains, lush public parks, tuition-free colleges, guaranteed jobs, shorter working hours and other items of 'carbon-free, communal luxury' (Aronoff et al., 2019, p. 172).

To a certain extent, the utopian hue of this vision is motivated by strategic considerations. According to the authors of *A Planet to Win*, an upbeat message is needed to 'bring labor on board'. Austere rhetoric should be avoided since no one is attracted by doom-and-gloom messages. 'Who will march for green austerity?', the authors ask rhetorically (Aronoff et al., 2019).

It is imperative to avoid the belt-tightening green politics of sacrifice. [...] More sacrifice to fix climate change is just not a winning political message, which is why a vision of public luxury and non-austere ways of living is important. (Battistoni & Cohen, 2020)

Significantly, not even radical variants of the Green New Deal negate the growth imperative of the capitalist economy, or capitalist production as such. The calls for 'system change' come with the caveat that a switch to a socialist mode of production is not possible within the brief time span at our disposal for a transition to a society free from fossil fuel. Abolishing capitalism will have to wait since the climate must be saved first (Aronoff et al., 2019, pp. 5, 19). The Green New Deal's affinity to liberal or mainstream variants of the green progress narrative is also evident in the pragmatic readiness of some of its proponents to affirm nuclear power and other large-scale technological solutions (ibid., p. 28).[2] Despite the socialist point of departure and the call for mass mobilization, there is thus a heavy reliance on technical solutions and the possibility of saving the climate without fundamentally challenging the social relations of capitalism.

DEBATES AND CRITICISM

The green progress narrative, we have claimed, is characterized by an ambivalence that implies, on the one hand, a belief in the inevitability of human progress and, on the other, a belief that action is necessary for further progress. This ambivalence is reflected in the stance taken towards the present system, which is both criticized and defended, as well as in the role that the environmental movement assigns to itself, which oscillates between leading the way into the future and affirming historical tendencies already in operation.

Connected to this ambivalence is the peculiar temporality of progress, which implies both continuous movement towards the new and remaining the same. Despite their differences, the liberal and the socialist variants of the green progress narrative are united by an underlying trust in history as a benign force that will push us in the right direction. Even where agency

---

[2] See the contributions to *Jacobin*'s special issue in 2017, 'Earth, Wind and Fire' (*Jacobin*, 15 August 2017, http://www.jacobin.com/issue/earth-wind-and-fire) in which authors like Leigh Philips, Peter Frase, and Christian Parenti ridicule catastrophism and 'hairshirt austerity' and endorse ecomodernist solutions in line with the Breakthrough Institute, e.g. nuclear power, geo-engineering, and carbon-capture technologies.

and action are given an important role, history is seen as providing favourable conditions for such action. Not even the most system-critical, socialist variants of the narrative call for a radical break with the status quo, but rather for a continuation and furthering of praiseworthy tendencies, above all the development of technology, that are already under way. Criticism is reserved for forces that obstruct or fetter these tendencies.

It is hardly surprising that critics of the green progress narrative have questioned its underlying trust in the tendencies of the present system, as well as the elements of historical determinism that underpin it. If newness is never more than an improvement of what already exists, then what if the system is fundamentally flawed? What if environmental disasters are the fruits of progress itself, rather than mere mishaps along the way, as argued by early critics of the progress narrative such as Max Horkheimer and Theodor Adorno (2002) or Lewis Mumford (1961)? These questions form the very core animating the present-day debate on the Anthropocene, since they concern whether human control over the planet should be seen as a source of anxiety and dread or as an opportunity for realizing a 'green' utopia with the help of science and technology (Hamilton, 2010). As critics have pointed out, the large-scale technologies that are often hailed as potential planet-savers—geo-engineering, nuclear power, carbon capture and storage, and so on—all have potentially catastrophic consequences (e.g. Vettesee & Pendergrass, 2022). While wind and solar power may seem more benign, these technologies have been criticized too. Not only does it seem unlikely that they will ever be able to fully replace fossil fuels and at the required speed, but there is also a clear risk that heavily relying on them in decarbonizing the economy will fuel an increasing number of 'green-green' conflicts, with conflicts around mining, land use, and wildlife habitats being the most conspicuous ones.

Apart from the *dangers* and *risks* of the new technologies on which the green progress narrative has often pinned its hopes, relying on them has also an *ideological* or *depoliticizing* function that has often been the target of incisive criticism. When the narrative stresses the inevitability of progress or affirms capitalism and the political system as well-functioning, it devalues protests from those who have been the victims of such progress. The idolization of science, technology, capitalist markets, and economic growth can easily appear cynical in view of the role they have played not only in the exploitation of nature but also in imperialism, widening inequality, and other forms of injustice (Hornborg, 2011). These injustices may well continue even in a 'green' economy, as seen in cases when the rights of indigenous peoples are sacrificed since their land is needed for mining or building wind parks. These wider social questions indicate the limitations of the narrow focus on natural beauty, recreation or scientific value that dominated the preservationist tendency within the early conservation movement, as well as the efforts to promote that prudent use of natural resources that came to dominate conservationist thinking and which later developed into today's idea of 'sustainable development'.

While Leftist versions of the green progress narrative give a more central place to the social nature interest, they too have been criticized for a naïve reliance on economic growth and technological advancement. In connection with the general criticism of the technological utopianism of the 'accelerationist Left' (Gardiner, 2017; Gordon, 2022; Noys, 2014) and of the pro-growth rhetoric of initiatives such as the Green New Deal (Foster, 2017), a seemingly interminable debate has raged about the environmental pros and cons of economic growth and the interlinked debate about whether a 'decoupling' of such growth from environmentally harmful production is possible (Hickel & Kallis, 2019; York, 2012). If the opening salvoes of this debate were fired with the publication in 1972 of the Club of Rome's report *The Limits to Growth* (to which we will return in the next chapter), it has gained renewed intensity in recent years with the rise of advocates of 'degrowth' within the environmental movement (Hickel, 2019, pp. 21f, 139–145; Kallis, 2018; Pollin, 2018; Stuart et al., 2021, p. 65f). As critics have pointed out, it is virtually impossible to reduce carbon emissions at the pace required to prevent dangerous global warming as long as the economy continues to grow (Hickel, 2019, p. 21). The growth imperative is also a main factor behind many 'green-green' conflicts. Decarbonizing the economy while maintaining growth is clearly impossible without the massive investments in mining, wind parks, and so on that give rise to these conflicts, which appear insoluble as long as the growth imperative is taken for granted. Much of the behaviour promoted as 'sustainable' may in fact best be understood as ways to 'sustain the unsustainable' (Blühdorn, 2017). The seeming impossibility of 'sustainable development' invites the apocalyptic fear to which we will now turn.

References

Aronoff, K., Battistoni, A., Cohen, D. A., & Riofrancos, T. (2019). *A Planet to Win: Why We Need a Green New Deal*. Verso.

Bastani, A. (2019). *Fully Automated Luxury Communism: A Manifesto*. Verso.

Battistoni, A. & Cohen, D. A. (2020, March 11). The Return of the Green New Deal: Ecosocialism in the USA. *Green European Journal.* https://www. greeneuropeanjournal.eu/the-return-of-the-green-new-deal-ecosocialism-in-the-usa/

Beck, U. (1995). *Ecological Enlightenment.* Humanities Press.

Blühdorn, I. (2017). Post-Capitalism, Post-Growth, Postconsumerism? Eco-Political Hopes Beyond Sustainability. *Global Discourse, 7*(1), 42–61.

Cassegård, C., & Thörn, H. (2023). *I skuggan av apokalypsen: miljörörelser och industrikapitalism* (pp. 1870–2020). Daidalos.

Foster, J. B. (2017). The Long Ecological Revolution. *Monthly Review,* 69(6). https://monthlyreview.org/2017/11/01/the-long-ecological-revolution/

Fox, S. (1981). *The American Conservation Movement: John Muir and His Legacy.* University of Wisconsin.

Gardiner, M. E. (2017). Critique of Accelerationism. *Theory, culture & society, 34*(1), 29–52.

Gordon, P. (2022). Left Accelerationism, Transhumanism and the Dialectic: Three Manifestos. *New proposals, 12*(1), 140–154.

Grove, R. (1995). *Green Imperialism: Colonial Expansion, Tropical Island Edens and the Origins of Environmentalism, 1600–1860.* Cambridge University Press.

Hamilton, C. (2010). *Requiem for a Species: Why We Resist the Truth about Climate Change.* Earthscan.

Haraldsson, D. (1987). *Skydda vår natur!: Svenska naturskyddsföreningens framväxt och tidiga utveckling.* Lund Univ. Press.

Hickel, J. (2019). *Less is More: How Degrowth Will Save the World.* Windmill.

Hickel, J., & Kallis, G. (2019). Is Green Growth Possible? *New Political Economy, 25*(4), 469–486.

Hisai, E. (2018). Whose Life Should Be Reformed?: The Transformation of the Life Reform Movement in Prewar Japan. *Asian culture and history, 10*(2), 10–18.

Horkheimer, M., & Adorno, T. W. (2002). *Dialectic of Enlightenment: Philosophical Fragments.* Stanford University Press.

Hornborg, A. (2011). *Global Ecology and Unequal Exchange: Fetishism in a Zero-Sum World.* Routledge.

Kallis, G. (2018). Degrowth. *Agenda publishing.*

Klein, N. (2014). *This Changes Everything: Capitalism vs. the Climate.* Simon & Schuster.

Klein, N. (2020). *On Fire: The Burning Case for a Green New Deal.* Penguin Books.

Krabbe, W. R. (1974). *Gesellschaftsveränderung durch Lebensreform.* Göttingen.

Krabbe, W. R. (1998). Lebensreform/Selbstreform. In D. Kerbs & J. Reulecke (Eds.), *Handbuch dere deutschen Reformbewegiungen 1880–1993.* Wuppertal.

Lenin, V. I. (2012). *Collected Works, Vol 31.* April-December 1920. https://www. marxists.org/archive/lenin/works/cw/pdf/lenin-cw-vol-31.pdf.

Merchant, C. (2004). *Reinventing Eden: The Fate of Nature in Western Culture.* Routledge.

Mol, A. P.-J., & Spaargaren, G. (2000). Ecological modernisation Theory in Debate: A Review. *Environmental Politics, 9*(1), 17–49.

Mumford, L. (1961). *The City in History: Its Origins, Its Transformations, and Its Prospects.* Harcourt.

Noys, B. (2014). *Malign Velocities: Accelerationism and Capitalism.* Zero books.

Polanyi, K. (2001). *The Great Transformation: The Political and Economic Origins of Our Time.* Boston: Beacon Press.

Pollin, R. (2018). De-growth vs a Green New Deal. *New Left Review, 112,* 5–25.

Sharma, A. (2012). Wilhelmine nature: natural lifestyle and practical politics in the German Life-reform movement (1890–1914). *Social History, 37*(1), 36–54.

Stolare, M. (2003). *Kultur och natur: moderniseringskritiska rörelser i Sverige 1900–1920.* University of Gothenburg.

Stuart, D., Gunderson, R., & Petersen, B. (2021). *The Degrowth Alternative: A Path to Address our Environmental Crisis?* Routledge.

Sveriges Natur (1913). Stockholm: Svenska Naturskyddsföreningen.

Thörn, H. (1997). *Rörelser i det moderna. Politik, modernitet och kollektiv identitet i Europa 1789–1989.* Stockholm: Rabénförlagen.

Thörn, H. (1999). "Nya sociala rörelser och politikens globalisering: demokrati utanför parlamentet?", in E. Amnå (ed.) *Civilsamhället (p. 84).* Stockholm: SOU.

Thörn, H., & Svenberg, S. (2016). 'We Feel the Responsibility That You Shirk': Movement Institutionalization, the Politics of Responsibility and the Case of the Swedish Environmental Movement. *Social Movement Studies, 15*(6), 593–609.

Thörn, H. (2019). Revolution as a politics of time-space: From Enlightenment Modernity to Advanced Globality. In Namli, E. (ed) *Future(s) of the Revolution and the Reformation* (pp. 65–95). Basingstoke: Palgrave Macmillan.

Treitel, C. (2017). Eating Nature in Modern Germany. In *Eating Nature in Modern Germany: Food, Agriculture and Environment, c.1870 to 2000.* Cambridge University Press.

Vettesee, T., & Pendergrass, D. (2022). *Half-Earth Socialism.* Verso.

Wapner, P. (2010). *Living through the end of nature: the future of American environmentalism.* MIT Press.

WCED, World Commission on Environment and Development. (1987). *Report of the World Commission on Environment and Development: Our Common Future.* Oxford University Press.

Williams, A., & Srnicek, N. (2013, May 13). #ACCELERATE MANIFESTO for an Accelerationist Politics. *Critical Legal Thinking.* https://criticallegalthinking.com/2013/05/14/accelerate-manifesto-for-an-accelerationist-politics/

York, R. (2012). Do alternative energy sources displace fossil fuels? *Nature, 2*(6), 441–445.

CHAPTER 3

Apocalypse

Abstract This chapter focuses on the apocalyptic narrative, which became central to the environmental movement after World War II. Fear of a threatening future is the pivot around which this narrative turns. Fear, however, is combined with hope that catastrophe can still be averted. Since it is deployed to mobilize action, the narrative is both action-oriented and future-oriented. In the 2000s, the apocalyptic rhetoric was adopted by mainstream actors who used it to bolster support for technocratic or market-oriented solutions. This contributed to increasing criticism against the narrative from scholars and activists who pointed out that its invocation of a common humanity blurred questions of inequality and responsibility and that the apocalypse was already a reality for large parts of humanity.

Keywords Environmental movement • Narrative • Apocalypse • Catastrophe • Doomsday debate

INTRODUCTION

Like the narrative of green progress, the apocalyptic narrative has deep roots in Western culture. It may even be argued that, in the context of Western modernity and its linear concept of time, the apocalypse is the dark sibling of progress. Ever since the early modern period and up until

© The Author(s), under exclusive license to Springer Nature Switzerland AG 2022
C. Cassegård, H. Thörn, *Post-Apocalyptic Environmentalism*,
https://doi.org/10.1007/978-3-031-13203-2_3

the twenty-first century, millenaristic (or chiliastic) movements have regularly appeared, declaring that an apocalypse is nigh. Strikingly often they have been offsprings of the legacy of the Protestant Reformation, its leaders' charismatic speeches borrowing heavily from the Book of Revelation. In the sixteenth-century German peasant uprising led by Thomas Müntzer (Scott, 1989) and the English seventeenth-century Levellers/Diggers (Hill, 1972), an ideal of social justice was part and parcel of the apocalyptic narrative. This narrative involved an ambivalence, which has tended to reappear in later versions of apocalyptic movement narratives. On the one hand, the apocalypse brings the world as we know it to an end; on the other, it gives birth to a new Millennium in which the chosen few enter into an earthly paradise, defined by peace, justice, and plenty.

When secularized versions of the apocalyptic narrative began to appear in the late nineteenth- and early twentieth-century socialist revolutionary movements, this ambivalence was rearticulated as an either/or: *unless* we succeed in revolutionizing the prevailing order, civilization will collapse. For example, Rosa Luxemburg (1971, p. 367f) in 1918 paraphrased both the Bible and the Communist Manifesto: 'The words of the **Communist Manifesto** flare like a fiery *menetekel* above the crumbling bastions of capitalist society: Socialism or barbarism!'

In this sense, the utopian action forms that define modern social movements are the children of both the early modern millenaristic movements and the eighteenth- and early nineteenth-century revolutions (see Chap. 1). As constructed through action-oriented movement texts, the narrative of progress and that of apocalypse both mobilize hope. But unlike the narrative of progress, in which hope is inextricably linked to the imagination of a better future as the main theme, the mobilizing emotion of the apocalyptic narrative is fear, linked to its main theme of the future as disaster. Hope can emerge only in a second step, in the process of collective action. This chapter focuses mainly on the emergence of the apocalyptic narrative in the decades following the Second World War; beginning with the so-called 'doomsday debate', Rachel Carson's *Silent Spring*, and Barry Commoner's *Closing the Circle*, all of which were highly influential when the environmental movement was established across the globe in the early 1970s. We take a closer look at a few of the early apocalyptic articulations in this movement context before moving on to consider the early 2000s, when the apocalyptic narrative gained new strength in light of the urgency of the issue of climate change.

From Extinction to Apocalypse

An apocalyptic narrative appeared already in the margins of the late nineteenth- and twentieth-century nature conservation movement. For example, in 1915, the Swedish conservation movement's journal *Swedish Nature* carried an article by Pehr Arvid Säve, previously published in 1877. The journal saw fit to republish the article on account of its prophetic qualities. In the article, Säve stated that 'the destruction of life' was a process that was on-going 'almost everywhere on the planet' (Säve, 1915, p. 4, our translation). However, it would perhaps be more accurate to name this a narrative of *extinction*, emphasizing a sense of loss rather than one of fear (of facing an apocalypse). As this Swedish example shows, the early environmental movement did not envision an imminent universal apocalypse that would annihilate all living beings on the planet. When terms such as 'extinction' or 'extermination' occurred in movement texts produced by the nature conservation movement, it most often referred to how modern civilization's rapidly expanding colonization of nature either threatened to bring about, or had already completed, the extinction of animal species (the dodo and the wisent being symbol-laden examples). It was thus not humans or societies that were under threat of extinction, but 'wild nature'. In the romantic concept of nature, this included animals and, as discussed in Chap. 2, also indigenous peoples, who were seen as part of nature, or at least as standing closer to nature than to 'civilization'. In this position they were, in some cases, including the North American indigenous peoples, the Scandinavian Sámi or the Australian aboriginals, regarded as endangered species.

In terms of praxis, the defence of endangered species was often anchored in a *scientific nature interest* in the sense that the measures taken to prevent the disappearance of animals, wild flowers or indigenous peoples were motivated by the argument that such an extinction would deprive humans of a source of knowledge about the planet's (natural) history. However, such defence could also be anchored in an *aesthetic nature interest*; as for example in those nationalist discourses where nature was conceived of as an important part of what defined the uniqueness of the nation's soul (as, for example, in Japanese, Swedish or North American nationalism). Further, from the perspective of a *recreative nature interest*, which also played an important role in early conservationism, the destruction of nature was perceived as a threat to (national) public health.

The Apocalypse as a (Post-)war Experience

The apocalyptic narrative was not only embraced by sections of the environmental, communist, and socialist movements in the first half of the twentieth century. The two world wars and the warmongering that preceded them provided fertile ground for the apocalyptic imagination. Particularly in the interwar period, prophesies of doom proliferated in light of the possibility of a second world war in the near future, and the probability that it would bring mass death on a scale that by far exceeded that which had occurred in the very recent past. Different prophesies of doom were articulated both by those who protested war and those who glorified it, albeit emphasizing different sides of the apocalyptic narrative's ambivalent future horizon. The German Nazi movement, which articulated the most elaborated fascist apocalyptic narrative (through movement intellectuals such as Alfred Rosenberg) presented the creation of the Third Reich as the dawn of a new Millennium that would restore a glorious European past; while the peace movement emphasized how this dark vision would bring death and destruction on a scale unprecedented in human history.

There is no doubt that the early post-war environmental apocalyptic narrative was articulated in the shadow of the nuclear disasters of Hiroshima and Nagasaki and the continued nuclear rearmament defining the Cold War. Not only did the post-war environmental movement emerge out of the anti-nuclear movement, but the early texts of the new environmentalism also consciously invoked the image of the nuclear mushroom cloud to convey the message about the coming of a new age in which environmental destruction occurred on an entirely new spatial scale. Environmental disasters were known previously, but only as local phenomena. From Engels's critique of the disastrous effects of industrial capitalism on the environment in *The Condition of the Working Class in England* (published 1844), to conservationists' and preservationists' conservative critique of the human destruction of nature in the early twentieth century, environmental destruction had mostly been understood as a local phenomenon.

In post-war environmentalism, the image of the exploding nuclear bomb became an image, comprehensible to anyone, of how industrial civilization had created technologies that threatened the atmosphere and nature as global systems. For example, in the words of Rachel Carson's *Silent Spring*: 'Along with the possibility of the extinction of mankind by nuclear war, the central problem of our age has therefore become the contamination of man's total environment [...]' (Carson, 1964, p. 13). In *The Closing Circle*, Barry Commoner explains that it was the nuclear tests of the 1950s that provided him with insights into how 'the "progress" of modern civilization is only a thin cloak for a global catastrophe' (Commoner, 1971, p. 298). And when Georg Borgström argued that science could be in service of human progress, he pointed out that this would require a radical change in the direction of its use so far: 'A common battle against starvation, disease, and misery, and above all against ignorance, requires a radical change in the goals of world science' (Borgström, 1965, p. 47). In the following section, we will take a closer look at the apocalyptic imaginations of these three scientists and movement intellectuals.

THE DOOMSDAY DEBATE

The nuclear bomb did not just provide a new, material foundation to the critique of human mastery over nature through science that had been an undercurrent throughout modernity. The nuclear mushroom also became a symbol for the combined threats of overpopulation and environmental destruction that defined the so-called 'doomsday debate'. This debate emerged under the influence of, and was sustained by, a few highly influential books, most of them written by natural scientists in the United States, including William Vogt's *Road to Survival* (published 1948), Fairfield Osborn's *Our Plundered Planet* (published 1948), Georg Borgström's *The Hungry Planet* (published 1965), and Paul Ehrlich's *The Population Bomb* (published 1968). As these authors' main focus concerned the risk of overpopulation, they have been labelled neo-Malthusians (e.g. Linnér, 2003). What was new about the neo-Malthusian apocalyptic discourse was that it brought an ecological perspective to Thomas Malthus's theory of how increased food production will increase the population, which in turn will increase demand for food until ultimately the earth is drained. The doomsday discourse added over-exploitation of other natural resources as well as several additional threats to the ecological system, such as nuclear war and the pollution of air and water. These were the core elements of the new apocalyptic narrative that was established in the Global North in the decades following the Second World War. It also emphasized the globality of the ecological system, and how it had been profoundly transformed by human action, thus implying a concept of second nature. Taken together, the numerous potential disasters, such as water shortages or desertification, provided good reasons to fear an ecological system collapse. Thus, for the first time, the neo-Malthusians argued, fears of a global apocalypse are substantiated by science. This strand of neo-Malthusianism culminated in 1972 with the publication of *Limits to Growth* by the Club of Rome, published just before the UN's first global environmental conference in Stockholm in 1972. Based on computer simulations, *Limits to Growth* dated the global collapse to 2072 (though it estimated that food production would begin to fall already in 2020).

In the works of Vogt, Borgström, and Ehrlich, the nuclear mushroom graphically illustrated the 'population explosion' that occurred at the time of the global breakthrough of industrial capitalism. For example, Borgström's illustration in a 1964 book estimates that a population growth of approximately 250 million per year since the time of year 0 in the Western calendar remained relatively steady until 1900, when the explosion occurred. From this point the 'dark cloud' of population growth expands, leading to a prognosticated global population rate of 4 billion in 1975 (Borgström, 1964, p. 14). To Borgström, this figure was catastrophic. While Vogt and Osborn founded this new apocalyptic genre, Borgström's and Ehrlich's books were timelier as they tied into the environmental discourse of the 1960s that give birth to the new environmental movement. Ehrlich's *Population Bomb* was the most successful and even sold more copies than Carson's *Silent Spring* (Linnér, 2003, p. 172).

Like the narrative of green progress, the new apocalyptic narrative could harbour different and sometimes contradictory political discourses and action strategies. When Paul Ehrlich was interviewed in *Look* magazine in connection with the first Earth Day in 1970, he accounted for his personal and highly political conclusions drawn from his work on *The Population Bomb*: 'When you reach a point where your further efforts will be futile, you may as well look after yourself and your friends and enjoy what little time you have left'. And he added: 'With present population growth that point for me is 1972' (quoted from Linnér, 2003, p. 172). These quotations show how one version of the apocalyptic narrative is embedded in a bourgeois sensibility, captured in a popular sentence that is often associated with Fredric Jameson (2003, p. 76): 'it's easier to imagine the end of the world than the end of capitalism'. Confronted with a profound insight about the possible catastrophic consequences of capitalism, the bourgeois subject's refusal, or inability, to imagine an alternative to capitalism, leads to resignation and passivity. The historian Christopher Lewis (1991) has also pointed to how the North American reception of

Vogt's *Road to Survival* tended to focus on the apocalyptic message rather than on how Vogt emphasized the contradiction between industrial capitalism and the environment, and advocated a revolution to establish a decentralized social organization in the spirit of Kropotkin (Linnér, 2003, p. 37).

In contrast to Ehrlich, Borgström's apocalyptic narrative included an emphasis on political action, linked to a social nature interest. Even though Borgström himself had national and supra-national institutions in mind, this is probably what made him relevant to the emerging environmental movement. For Borgström, it was evident that the apocalypse could be avoided should the right political measures be taken, and these must include a global redistribution of resources. At the same time, his books are riven by deep ambivalence regarding the possibilities to steer the planet away from disaster. In particular, the Preface to *The Hungry Planet* paints a dark picture: 'As things now stand we seem to face the alternative of nuclear annihilation or universal suffocation' (Borgström, 1965, p. viii). He then borrows an image from the film *Modern Times*, to illustrate the futility of our pursuit for progress: Chaplin tries to run up an escalator as it continues to move downward, just as 'our great human escalator is persistently accelerating its downward movement' (ibid., p. ix). Yet, there 'still remains the choice' for humankind to take a different path, but the time is 'five minutes to twelve' (ibid., p. viii). This ambivalence was shared by the movement intellectuals of the emerging new environmental movement. It can even be argued that such ambivalence is a prerequisite for the mobilization potential of the apocalyptic narrative; in order to move people to act to avoid disaster, the threat of an apocalypse must appear as real, in the sense of it being a real possibility. Since pre-modern times, doomsday prophets acquired their authority from the Bible when painting such dark visions. This time they were backed by science. However, Borgström, at the time Professor of Food Science at Michigan State University, also drew from the Bible to convince his readers that scientific resources need to be mobilized to save the planet rather than taking its population into space:

> The holy writ states: 'For what does it profit a man if he gains the whole world and forfeits his soul?' A paraphrase of this this biblical text, fit for our time and age would be: 'What does it profit man if he gains the universe but loses the earth?'. (Borgström, 1965, p. 470)

And it is precisely this process that is under way, according to Borgström: 'we are being robbed of our biological environment' (ibid., p. 457). The second nature that replaces it is an 'artificial, supertechnical world' (ibid., p. 457), produced by scientific reason. However, as Borgström emphasizes, unlike the Bible, the Book of Science is, as a Pharmakon, both poison and cure at the same time, meaning that there is no absolute authority to guide the future destiny of humanity. Thus, the ambivalence as to whether catastrophe can be avoided is inextricably linked to the ambiguous role of science; it helped produce environmental destruction, it diagnosed it, and it may yet help in averting disaster. For the latter to happen, however, modern science must take a completely new direction. To make evident how misguided the efforts of science have become, Borgström adopts the perspective of an imagined future looking back on our present:

> The historians of the future will discover—if archival material covering this development is available—to their astonishment the distortion between the funds which in our age were allocated to areas like space and atomic research and those for biology and medicine. (ibid., p. 466)

Borgström continues in an ironic mode to carry out a thought experiment: if space research were really the solution to the world's most urgent problem, 'we would be faced with the overwhelming problem of sending away in spaceships more than 7500 persons per hour, around the clock, month after month, year after year' (ibid., p. 470).

In *The Hungry Planet*, Borgström uses the concept of 'ghost acreage', his most important contribution to environmental as well as social science (anticipating the idea of ecological footprint). This concept was designed to illustrate the 'invisible' use of resources, which is a condition for the production of consumer products in the richest parts of the world, but which is excluded from dominant economic models.[1] With this notion, Borgström made clear how the rich countries, through unequal exchange, continued to exploit previously colonized countries in the Global South even after these countries had gained political independence. In a later

---

[1] Borgström (1965, p. 71) discerns two elements of ghost acreage. The first element is 'the computed, non-visible acreage which a country would require as a supplement to its present visible agricultural acreage in the form of tilled land in order to be able to feed itself'. The second element is 'represented by *trade acreage—calculated as the acreage, in terms of tilled land, required to produce, also with present techniques* the agricultural products constituting the *net importation*'.

book, using a computer model, Borgström estimated that Western Europe's consumption of resources that originated from other continents corresponded to a cultivation area equal to all cultivated land in Latin America (Borgström, 1975, p. 36). For Borgström, this was also proof for his argument that the current level of development is unsustainable; should the entire world population be elevated to the standard of Western Europe, it would signal ecological collapse. Thus, Borgström's articulation of a social nature interest involves a global perspective on the distribution, exchange, and management of natural resources.

As a son of a preacher, Borgström mastered the apocalyptic language. There are also elements of a postapocalyptic narrative in his diagnosis of the conditions in which most of the world's population live, that is, the starving and undernourished populations in the Global South. For such people, the catastrophe is already a reality (Borgström, 1964). At the beginning of the 1970s, when his pessimism grew darker based on two decades of sounding the alarm, this dimension became more accentuated. When he concluded *Focal Points: A Global Food Strategy* (1971), the time was no longer five minutes to twelve: 'The time is five minutes past twelve' (Borgström, 1973, p. 258).

'The Rumblings of an Avalanche'

With the popularity in the 1960s and early 1970s of the works by neo-Malthusians such as Paul Ehrlich and Georg Borgström, and environmentalists such as Rachel Carson and Barry Commoner, the critique of human mastery over nature through science began to surface as 'common sense'. The nineteenth-century secularized idea of human development through evolutionary jumps that moved through ever higher stages that Herbert Spencer, Karl Marx, and Emile Durkheim translated into radically different versions of social science, no longer seemed self-evident to a secularized modern subject. Marx (1976, p. 461) had referred to Darwin's *On the Origin of Species* as an 'epoch-making work' in *Capital, Volume One* and he obviously felt that there was an affinity between his own work and the natural scientists; they were both engaged in a scientific inquiry into the laws of natural and human history, respectively; and the spectre in the *Communist Manifesto* was a playful metaphor for the role of human agency, rather than extra-social forces, in such evolution. In comparison, Carson's *Silent Spring* appears as an inverted image of the faith in science, progressive human evolution, and human agency, which defined Marx's work. The first chapter of *Silent Spring*, 'A Fable for Tomorrow' is a fictional story that begins like a fairy tale about a green and stunningly beautiful town in 'the heart of America', but the story ends like a tragedy when a series of disasters hit the town, causing the sudden death of humans and animals. Like Marx and Borgström, Carson uses the metaphor of the spectre to signify a force that in a sense is hidden beneath the surface of the order of things: 'A grim spectre has crept upon us almost unnoticed, and this imagined tragedy may easily become a stark reality we all shall know' (Carson, 1964, p. 11).

There is nothing mysterious or supernatural about this spectre. As in the *Communist Manifesto*, the spectre here represents human agency bringing about profound social change; however, 'the people', since the late eighteenth century invested with hope by progressives, is here bringing death and destruction: 'No witchcraft, no enemy action had silenced the rebirth of new life in this stricken world. The people had done it themselves' (ibid., p. 10). Like Marx, Carson also invokes one of her own century's most highly regarded intellectuals to provide authority to her narrative. On the first page of the book there is a dedication to, and quotation from, Albert Schweitzer, winner of the 1952 Nobel Peace Prize. While Schweitzer, like Darwin, was a natural scientist, he was first and foremost a philosopher critiquing modern civilization: 'Man has lost the capacity to foresee and to forestall. He will end by destroying the earth' (ibid., p. 5).

Overall, time is a fundamental theme in the critique of modern society developed by Carson in *Silent Spring*. Most important, she establishes a narrative strategy that would define the apocalyptic narrative embraced by the new environmental movement; the present is regarded from the viewpoint of a future in which the catastrophe has become a *fait accompli*. Importantly, Carson emphasizes that the future is not entirely imagined, but is in a sense already here, in the present, since all the disasters that have struck the little town in Carson's fable 'actually happened somewhere, and many real communities have already suffered a substantial number of them' (ibid., p. 11). In this sense, while Carson's narrative is primarily apocalyptic, there are also elements of a postapocalyptic narrative.

The present time in Carson's book is the post-war era. In this short period, humanity has reversed the relation between living things and their environment that so far had defined the history of the planet. Here Carson's early formulation of an ecological nature interest is clear. Humanity has fundamentally disturbed the balance of the planetary ecosystem. While previously, life forms in most cases had been shaped by the environment, one species is now shaping the environment on a global scale. Carson is thus implying the concept of second nature as she argues that after the Second World War, humanity has, through scientific and technological development, 'acquired significant power to alter the nature of his world'. This power 'has not only increased to a disturbing magnitude'; its character has also changed so that we now see 'a universal contamination of the environment'. Moreover, a major part of this pollution is 'irrecoverable' (ibid., p. 12).

This is an example of the disastrous consequences of how humanity in the modern epoch has changed time. According to Carson, a fundamental condition for the adjustment of life to its environment in the history of the planet is what we would term evolutionary time, which is slow time, 'time not in years but in millennia'. In modernity, humanity has increased the speed of time to the point where it has been annihilated: 'in the modern world there is no time' (ibid., p. 12). This also entails a new relation between the present and the future. Through promises of radical progress articulated by the scientists posing as the 'architects of our future' (ibid., p. 13), we could experience the future as progress. However, the future has struck back; instead of the present colonizing the future in the shape of an ever-advancing scientific progress at increasing speed, the future has colonized the present in the shape of risks of global disaster. While Carson points to how humanity, through the destruction of its environment, has become its own enemy, the major responsibility for the destruction of nature that defines the present rests with a megalomanic science, which has been blind to the fact that it endangers life on earth with experiments that introduce 'substances that accumulate in the tissues of plants and animals and even penetrate the germ cells to shatter or alter the very material of heredity upon which the shape of the future depends' (ibid., p. 13). Nevertheless, Carson frequently quotes science against itself, making the modern scientist, just like humanity, an ambivalent figure in her narrative, potentially friend and foe.

Like Borgström's work, Carson's text carries a deep ambivalence regarding the possibility of retaking control over the present as well as the future, and of steering the planet in a direction that would avoid disaster. On the one hand, as the title of one chapter reads, we hear 'the rumblings of an avalanche'. But on the other, Carson's book could not have been adopted as a movement text a decade later were it not for the fact that it was a call to action. Carson's 'fable for tomorrow' is intended to alter our idea of the future as a promise of a better and higher form of life. Thus, the final chapter points to 'the other road'. Carson articulates the either/ or of the apocalyptic narrative by providing an image in which humanity is standing at a crossroads: 'We stand now where two roads diverge'. Thus, to Carson, the future is still open and dependent on human agency. Yet, collective action is not mobilized by utopian promises, but rather by the threat of a profound disaster. And the only way to avoid the latter is to give up 'control of nature' (ibid., p. 153) and adjust our societies to nature, rather than subordinate nature to society. The utopia of Carson's apocalyptic narrative thus also underscores her articulation of an ecological nature interest: humanity as a 'we' is a species that is a fully integrated part of a planetary system, which we must respect rather than dominate.

Closing the Circle

While both Carson's and Borgström's books presaged and helped to ignite the new environmental consciousness that gave birth to the post-war environmental movement in the US, Commoner's *The Closing Circle* was published a decade later, at a time when the movement started to emerge as a new and global phenomenon. This book made biologist Barry Commoner one of the leading intellectuals in the new environmental movement in the 1970s. Indeed, the 'four laws of ecology' he presented in *The Closing Circle* were adopted into the 1976 Greenpeace manifesto (see further below). Commoner began his activities as a movement intellectual more than a decade earlier in the context of the anti-nuclear movement, and in *The Closing Circle* he states that it was the nuclear bomb tests that provided him with his insights about 'the meaning of the environment and its importance to human life' (Commoner, 1971, p. 52). This may be interpreted as an experience of second nature inherent in the contemporary concept of the Anthropocene: the irreversible change of nature produced through nuclear debris reveals both the fundamental dependency of humanity on its environment, and how profoundly human beings have transformed it. To Commoner, this fundamental, man-made transformation of nature means that in terms of knowledge about nature, direct everyday experience of nature is now more limited than before:

> In the woods of Walden Pond or on the reaches of the Mississippi, most of the information needed to understand the natural world can be gained by

personal experience. In the world of nuclear bombs, smog, and foul water, environmental understanding needs help from the scientist. (ibid., p. 41)

While Commoner, like Carson, criticizes the scientific practices that have brought us what he calls 'ecological disaster' (ibid., p. 47), he is much more concerned with defending the science without which we would have no knowledge about 'what we recently have learned to call ecology' (ibid., p. 47). For Commoner, science needs to guard its autonomy, integrity, as well as the ethical principles that this involves in relation to those political and economic forces that work to employ scientific knowledge to dominate, exploit, and destroy nature. This was the main theme in his early book *Science and Survival* (1963), the first chapter of which is titled 'Are we losing control over science?' The third edition has a new introduction, which begins by linking the end of a blind belief in science to an auspice about a possible collapse of modern civilization:

> The age of innocent faith in science and technology may be over. We were given a spectacular signal of this change on a night in November 1965. On that night all electric power in an 80,000-square-mile area of the northeastern United States and Canada failed. (Commoner, 1966, p. 3)

Commoner also cites a report from the incident, stating that New York came 'to the brink of chaos' (ibid., p. 5). This sudden breakdown of a society, displaying the vulnerability of the dependence on complex technological systems, is one of two apocalyptic scenarios that Commoner presents in the book. We might call it *the sudden apocalypse*. In contrast to this, Commoner presents another apocalyptic scenario. Just a month after the electricity breakdown, nine children were taken to hospital with mysterious and identical symptoms. It turned out that they had thyroid cancer, and that the cause of this was that they had been exposed to radioactive iodine produced by fallout from atomic bomb tests in Nevada (ibid., p. 6). For Commoner, this is but one of many examples of a slow-acting poisoning of the ecological system, including the human body, on a global scale. We might call this scenario *the slow apocalypse*.

Considering these apocalyptic scenarios, the limited future goal that humanity could hope to realize is 'To survive on the Earth', as the title of the last chapter of *Science and Survival* reads. This is an early articulation of the alternative future envisioned by the new environmental movement when it emerged in the early 1970s; instead of imagining the future in terms of expanding growth and wealth, it adopted the utopia of *survival*. As indicated by the title of Commoner's book, science has a key part to play in realizing this utopia. But in essence, the task is political. Unlike Carson, who does not discuss the role of politics in *Silent Spring*, and distinct from Borgström, who placed his faith in established politics on the national and supra-national level, Commoner assigns a role to the environmental movement. The second edition of *The Closing Circle* (1971) starts with observations of activism during Earth Day in the US in 1970. Commoner is particularly interested in how different protest groups define the responsible subject of environmental destruction, including 'the capitalists', 'the politicians', and 'technology' (Commoner, 1971, p. 3). The observation is concluded by his own identification with someone who quotes a character in a comic strip: 'And one keen observer blamed everyone: We have met the enemy and he is us.—Pogo' (ibid., p. 4). Commoner thus seems to share Carson's position: humanity as such is both the enemy and the (only possible) subject of change. However, while Commoner clearly articulates an ecological nature interest, he departs from Carson in that he combines it with a social nature interest, which means that he differentiates the concept of humanity into social forces and classes, with differing relations to nature, as well as differentiated responsibilities for its destruction. This also means that for Commoner, the solution to the present crisis is not simply for humanity to adjust to an ecological system that works according to an extra-social logic; for him, the ecosphere also needs to be understood as a social phenomenon. In his dramatic version of the apocalyptic narrative's conception of the future as an either/or, Commoner argues that this has consequences for how we should imagine and indeed create a future society beyond the present environmental crisis:

> In the narrow options that are possible in a world gripped by environmental crisis, there is no apparent alternative between barbarism and the acceptance of the economic consequence of the ecological imperative—that the social, global nature of the ecosphere must determine a corresponding organization of the productive enterprises that depend on it. (Commoner, 1971, p. 297f.)

This means that 'as a political issue, environmental protection is neither innocuous nor unrelated to basic questions of social justice' (ibid., p. 207). Linking the environmental issue to structural racism, Commoner also anticipates the emergence of the environmental justice movement in the

US a decade later, as he argues that the particularly polluted environment of the ghetto makes black people more exposed to pollution than the white population: 'Blacks need the environmental movement, and the movement needs the blacks' (ibid., p. 208).

Just like Carson, Commoner's critique of the disastrous environmental destruction in modern society led him to question the linear concept of time inherent in the idea of progress. Commoner, however, took this critique a step further. While linear time is a delusion, real time is natural time, which is cyclical. Humanity has attempted to conquer nature by breaking out of this circle. The utopia of survival can therefore only be realised by a collective action through which we 'close the circle' (Commoner, 1971, p. 300).

Commoner was a transnational activist. In the 1960s, he was vice chairman of the Soil Association, established in 1945 by anthroposophists. Under the influence of Commoner and members of a younger generation of environmentalists, the Association, however, changed its orientation towards an ecological nature interest, and began to articulate a critique of the actions of multinational corporations (Bramwell, 1989). The Soil Association was also one of the organizations which (together with new transnational environmental organizations such as Friends of the Earth) were behind A Blueprint for Survival (1972). In a European context, this was the most important environmental movement manifesto of the early 1970s.

1972: The Birth of an Apocalyptic Global Environmental Movement

In 1972, the first global UN environmental conference was held in Stockholm. Swedish Prime Minister Olof Palme's inaugural speech to the conference testified both to the influence of the apocalyptic narrative and to an emerging popular environmental consciousness:

> Some years ago public discussion—at least in the industrial countries—centred around a probable future of affluence and abundance. It was not much disturbed by the reality of poverty and starvation for the vast majority of people in the world. [...] Nowadays, the debate largely centres around a future of scarcity on this one earth. Progress continues, yes, and world production increases. But we have become increasingly aware that our natural resources are limited. We have come to discuss more and more the

interrelated problem areas of population, poverty and pollution. [...] This has led to some prophecies of doom, and many expressions of gloom, concerning the future of the human race. It has led to a very genuine sense of fear among ordinary people all over the world. [...] People show concern and they demand action.

The concern of the ordinary people mentioned by Palme found expression in the alternative assembly held in connection to the UN meeting. In fact, 1972 may be seen as the year in which the post-war environmental movement started to establish itself in a number of countries across the world; for example, the first green party was founded in New Zealand. Further, as already mentioned, *A Blueprint for Survival* was published in England (a year before the first European green party was formed in the same country). The manifesto was originally published in *The Ecologist*, a journal (printed on recycled paper) that functioned as a forum for young green movement intellectuals in Britain. The manifesto cited recent, widely referenced sources that had contributed to establishing the apocalyptic narrative in the early 1970s, such as the Massachusetts Institute of Technology's *Man's Impact on the Global Environment: The Study of Critical Environmental Problems* (published 1970), and most important, *Limits to Growth* (published 1972). *A Blueprint* also cited Barry Commoner and Georg Borgström, thus contributing to the canonization of their books as movement texts. Yet, unlike these books, *A Blueprint* is *intended* as a movement text in the shape of a manifesto. It is tied to the foundation of the Movement for Survival which, according to the statutes, was a coalition of autonomous organizations, including Friends of the Earth and The Soil Association. The main aim of this new movement organization was to push governments (the British in particular) to take 'measures most likely to lead to the stabilisation and hence the survival of our society' (ibid., p. 23). Thus, in this utopia of survival, the concept of 'change'—one of the most important nodal points of modern movement discourse since the late eighteenth century—was replaced by 'stabilisation'. What is envisioned here is a future society which 'by definition would depend not on expansion but on stability' (ibid., p. 6). Further, in the articulation of the movement's key utopian action form, *A Blueprint* also introduced a concept that a decade later would be appropriated by the renewed narrative of green progress: 'creating a sustainable society' (ibid., p. 2). It was also clear that this action was guided by an ecological nature interest, since the utopian future will see 'the dawn of a new age in which Man will live with the rest of Nature rather than against it' (ibid., p. 1). However, different from the concept of sustainability that would later define the discourse of ecological modernization, this bright future needed to be seen as the (only possible) alternative to a global apocalypse:

> An examination of the relevant information available has impressed upon us the extreme gravity of the global situation today. For, if current trends are allowed to persist, the breakdown of society and the irreversible disruption of the life-support systems on this planet, possibly by the end of the century, certainly within the lifetime of our children, are inevitable. (ibid., p. 1)

The main task thus concerns 'removing the sword of Damocles which hangs over the heads of future generations' (ibid., p. 6). A significant difference in relation to Commoner's *Closing the Circle*, as well as Borgström's later texts, is that *A Blueprint* does not in any profound sense attempt to combine its ecological nature interest with a social one. It displays significant concern with global overpopulation, but it does not envision global redistribution as part of its construction of a sustainable society. Its view on inequality is instead national; it states that '[u]nless we are willing (and able) to perpetuate an even greater inequality of distribution than exists today, Britain must be self-supporting' (ibid., p. 13).

By comparing Commoner's vision with *A Blueprint*, we can get a sense of the tension between an ecological and a social nature interest that emerged at the very inception of the new environmental movement, and that would grow in the coming decades, up until the point where 'climate justice' was accepted as an umbrella term for the global climate movement (Cassegård & Thörn, 2017). In these battles each side would accuse the other of having an anthropocentric or biocentric bias, respectively. On the one hand, those advocating a combination of a social and an ecological nature interest, arguing that the fight against destruction and exploitation of nature was inextricably linked to the struggle for social justice, would be criticized for reproducing the anthropocentric view of the ecosystem that had taken humanity to the brink of destruction in the first place. On the other hand, the biocentrists would be criticized for a lack of insight into the fact that the ecosystem could not be separated from the social system, and that any attempt to create an ecologically sustainable society without addressing social inequalities was doomed.

The most radical biocentric wing of the new environmental movement may also be said to have been born in 1972. In this year, the Norwegian philosopher Arne Naess gave the inaugural speech at the 3rd World Future Research Conference, held in Bucharest. It was titled 'The shallow and the deep ecology movement'. A summary of the speech was published in *Inquiry* a year later (as 'The Shallow and the Deep, Long-Range Ecology Movement: A Summary') and subsequently became canonized both as a movement text and as one of the most influential articles in environmental ethics (Anker, 2008). In the article's introduction, Naess states that 'shallow' ecology is concerned only with pollution and resource depletion, while deeper concerns 'touch upon principles of diversity, complexity, autonomy, decentralization, symbiosis, egalitarianism, and classlessness' (Naess, 1973, p. 95). In his critical assessment of shallow ecology, Naess argues that its central objective is 'the health and affluence of people in the developed countries' (ibid., p. 95), and in this way he implies a social nature interest and recognizes that the conflict between rich and poor countries is one of the concerns of the deep ecology movement. However, the egalitarianism that Naess has in mind is 'biospherical', meaning that it includes all forms of life. This contrasts with what Naess argues is the anthropocentric view that restricts 'the equal right to live and blossom' (ibid., p. 96) to humans.

This radically biocentric perspective was also adopted by Greenpeace, established in Canada in 1971. Its 1976 manifesto, 'Greenpeace Declaration of Independence' states that 'mankind is not the centre of life on this planet'. The manifesto cites three of Commoner's laws of ecology which it argues 'hold true for all forms of life—fish, plants, insects, plankton, whales and humans'. If humans do not begin to adhere to these principles, it will lead to 'the destruction of the earth', and 'inevitably, to the destruction of ourselves'. Unlike Naess's manifesto, the justice theme is absent, and thus so too is any trace of a social nature interest. Further, in comparison to Naess, the Greenpeace Declaration explicitly ties its ecological nature interest to the apocalyptic narrative and its either/or, as its first sentence states:

> We have arrived at a place in history where decisive action must be taken to avoid a general environmental disaster. With nuclear reactors proliferating and over 900 species on the endangered list, there can be no further delay or our children will be denied their future. (Greenpeace, 1976)

In the 1970s, the apocalyptic narrative also became anchored in the protests against nuclear power stations that occurred in a number of countries in the Global North (e.g. Jasper, 1990; Flam, 1994). However, as mentioned in Chap. 2, as narratives of green progress enjoyed a resurgence in the 1980s and 1990s, the apocalyptic narrative was challenged as the dominant approach to future-oriented action in the environmental movement.

The apocalyptic narrative was, however, reinforced when climate change started to emerge as the main, overarching concern in the environmental movement during the first decade of the new millennium. While this new apocalyptic narrative thus centred around a new kind of disaster, its dramaturgy and tropes were basically similar to those that had mainly concerned overpopulation and the nuclear threat. Importantly, the issues of climate change, nuclear threat, and overpopulation were similar in the sense that fear of a collapse on a global scale was supported by substantial scientific evidence (though there was much less of a consensus around the issue of overpopulation compared to the other two). Like the nuclear threat, the climate issue could be linked to the two scenarios of slow and sudden apocalypse. While the former often involved a century-long perspective, in which the earth would become uninhabitable because of rising temperatures around the year 2100, the latter involved the risk of sudden collapse caused by so-called threshold effects.

News headlines in the run-up to the 2009 Copenhagen meeting, warning of a coming global collapse, as well as Al Gore's celebrated and highly influential film *An Inconvenient Truth*, indicated that the apocalyptic narrative was now becoming anchored in dominant institutions. At the same time, the climate justice movement entered the stage of global climate activism and brought an alternative apocalyptic narrative. It was different from the established one in the sense that feelings of anger overshadowed those of fear. For example, such a view was articulated by Swedish climate activist Andreas Malm, who a decade later, with the publication in 2021 of books such as *How to Blow Up a Pipeline* and *White Skin, Black Fuel*, would rise to become one of the leading radical climate movement intellectuals. In 2007, he published a book titled *It is our definite opinion that if nothing is done now it will be too late*. Along the lines of Rosa Luxemburg's either/or (cited above), Malm asserted that 'we have waited too long already, and the longer we wait, the more urgent it is that the attack on fossil fuel will be without compromise: a revolution' (Malm, 2007, p. 247, our translation).

An important difference between the apocalyptice narrative of Gore's film and that of the climate justice movement at the time, which also overlaps with the different views implied in the notions of Anthropocene /Capitalocene, concerns its enemy and its acting subject. In the narrative of the climate justice movement, it is not humanity, but capitalism, which has taken the planet to the brink of collapse, and neither is it humanity, but an alliance between climate science and the climate movement that represents the possibilities of avoiding such a disaster. A decade later, by the end of the 2010s, Extinction Rebellion (XR) and Fridays for Future engaged with the radical apocalyptic narrative and its ambivalent either/or. For example, in a speech held outside the Houses of Parliament in London in 2019, Greta Thunberg argued that 'we probably don't even have a future anymore', since it has 'been sold so that a small number of people could make unimaginable amounts of money' (Thunberg, 2019, p. 58). At the same time, Thunberg, like Rachel Carson 60 years earlier, identifies a space of action, claiming that 'humanity is now standing at a crossroads' (ibid., p. 56). However, Thunberg has also, like XR, oscillated between an apocalyptic and a postapocalyptic narrative, which is why we will return to them in the next chapter.

CRITIQUE OF THE APOCALYPTIC NARRATIVE

In this chapter, we have seen how the apocalyptic narrative in the context of the environmental movement is not anti-utopian, but seeks to mobilize through 'negative' utopian energies: the future as a threat of disaster and its either/or. Thus, survival could be understood as a utopia only in the context of an apocalyptic narrative. As a social movement narrative, the apocalyptic narrative is perhaps adequately characterized as a reluctant child of Enlightenment modernity in the sense that it promotes the idea that collective action can change current conditions and shape a new and better world. But in contrast to social movement narratives of Enlightenment modernity, apocalyptic discourse introduces an ambiguity in the idea of utopian action as future oriented. Like the pre-Enlightenment concept of revolution (Koselleck, 1985), apocalyptic utopia constructs a future that involves a cyclical concept of time, as the future is also a return to the past, in the sense of restoring a 'disturbed order'.

In the late twentieth century, the apocalyptic narrative's critique of progress, science, and linear time often provoked critics to dismiss it as anti-modern and quasi-religious. Given its new established framing in the early 2000s, a new kind of criticism emerged. For example, the geographer Erik Swyngedouw (2013) argues that the apocalyptic narrative had become part and parcel of a post-political framing of the climate supported by large segments of the establishment itself. The political scientist Chris Methmann (2013) similarly observes that alarmist reports invoking future catastrophes are insufficient to move politicians to take the necessary steps to curb the environmental destruction. The apocalyptic rhetoric, he states, is sustained by its success, not in reducing emissions, but rather, perversely, in depoliticization and deflecting criticism of the system. Further, in a Marxist take on this argument, Eddie Yuen points to the depoliticizing effects of the collective 'we' or 'everybody' in apocalyptic discourse. The abstract idea of 'Humanity', supposed to be both perpetrator and victim of environmental destruction, erases meaningful class and geographical differences—'Beware of plutocrats speaking of Spaceship Earth' (Yuen, 2012, p. 26). However, following the 2009 Copenhagen climate summit (COP15), which not only was a disappointment for those who had invested hope in a binding agreement but also created a deep split in the broader climate movement, a sense of crisis hit both established environmental organizations and the more radical networks that had mobilized around the concept of climate justice. In our research on the global climate movement in the years preceding the Paris 2015 COP meeting, we found that strategies, targets, and ideological foundations were questioned, discussed, and re-evaluated in all environmental movement camps. We also found that quite a few major organizations, as part of their re-strategizing post-Copenhagen, systematically discussed the role of narrative and emotions in processes of mobilization. One of the results of this broad re-strategizing process was the new climate movement unity, with climate justice serving as a discursive nodal point, that began in Warsaw in 2013 and that reached its apex during the 2015 Paris meeting (Cassegård & Thörn, 2017). Yet, this did not exclude internal friction and conflict, one of the most important of which related to different positions on how to transcend the apocalyptic imagery that had enjoyed a hegemonic position in the broad environmental movement for so long.

Among established environmental organizations in particular, the apocalyptic narrative was questioned from a strategic point of view; maybe the idea of articulating messages of a likely future disaster only instigated fear and created hopelessness and thus passivity, or perhaps it was potentially so psychologically burdensome that many people simply suppressed awareness of climate change? Should not movement discourses emphasize 'positive' feelings such as hope instead of 'negative' feelings such as fear? The result of these discussions seemed to be a widespread agreement that the environmental movement narratives should be shaped to emphasize positive feelings, particularly hope (and that it had previously overemphasized negative feelings, particularly fear) (Thörn et al., 2017).

From the camp that had mobilized around climate justice in Copenhagen, the apocalyptic discourse was criticized from a different angle: the whole idea of a future disaster was perceived as North-centric, considering that climate-related disasters were already occurring in many poor countries. To many people, especially in the Global South, the apocalypse is already here. Climate movement actors based in the Global South are thus one of the key contexts for the emergence of the postapocalyptic discourse that will be discussed in the next chapter. Swyngedouw (2013) connects with this South-based critique and argues it is this insight that is mobilizing. People are increasingly protesting not only out of fear for the future, but also out of anger at an already ongoing catastrophe and to demand justice.

We thus saw two emerging responses to the draining away of utopian energies from the apocalyptic discourse: a 'hopeful' one insisting on the positive rewards for action and our ability to avert catastrophic climate change if we act together; and a postapocalyptic one that takes its point of departure from the experience of irreversible or unavoidable loss. While the former arose out of strategic concerns about how to sustain effective mobilization, the question remains how postapocalyptic narratives can be deployed in political mobilizations. The answer, as we will argue below, is that mobilizations based on accepting loss are possible through what we call the paradox of hope and the paradox of justice.

References

A Blueprint for Survival. (1972). In *The Ecologist* (1), 1–22.

Anker, P. (2008). Deep Ecology in Bukarest. *The Trumpeter, 24*(1), 56–67.

Bramwell, A. (1989). *Ecology in the 20th Century: A History.* New Haven: Yale U.P.

Borgström, G. (1964). *Gränser för vår tillvaro.* LT.

Borgström, G. (1965). *The Hungry Planet: The Modern World at the Edge of Famine.* Macmillan.

Borgström, G. (1973). *Focal Points: A Global Food Strategy.* Macmillan.

Borgström, G. (1975). *Banketten.* Trevi.

Carson, R. (1964). *Silent Spring.* Fawcett Publications

Cassegård, C., & Thörn, H. (2017). Climate Justice, Equity and Movement Mobilization. In H. Thörn, C. Cassegård, L. Soneryd, & Å. Wettergren (Eds.), *Climate Action in a Globalizing World: Comparative Perspectives on Environmental Movements in the Global North* (pp. 33–56). Routledge.

Commoner, B. (1966). *Science and Survival*. Victor Galliancz.

Commoner, B. (1971). *The Closing Circle: Confronting the Environmental Crisis*. Jonathan Cape.

Flam, H. (1994). *States and Antri-nuclear Movements*. Edinburg: Edinburgh Press.

Greenpeace. (1976). Greenpeace Declaration of Independence. Downloaded May 31, 2022, from https://web.archive.org/web/20121008064009/http://rexweyler.com/greenpeace/greenpeace-history/declaration-of-interdependence/

Hill, C. (1972). *The World Turned Upside Down: Radical Ideas During the English Revolution*.

Jameson, F. (2003). Future City. *New Left Review, 21*, 65–79.

Jasper. J. (1990). *Nuclear Politics: Energy and the State in the United States, Sweden, and France*. Princeton, NJ: Princeton U.P.

Koselleck, R. (1985). *Futures Past: On the Semantics of Historical Time*. MIT Press.

Lewis, C. (1991). *Progress and Apocalypse: Science and the End of the Modern World*. Diss., University of Minnesota.

Linnér, B. (2003). *The Return of Malthus: Environmentalism and Post-War Population-Resource Crisis*. White Horse Press.

Luxemburg, R. (1971). What Does the Spartacus League Want? In D. Howard (Ed.), *Selected Political Writings of Rosa Luxemburg* (pp. 366–376). Monthly Review Press.

Malm, A. (2007). *Det är vår bestämda uppfattning att om ingenting görs nu kommer det att vara för sent*. Atlas.

Marx, K. (1976). *Capital* (Vol. 1). Penguin.

Methmann, C. P. (2013). The Sky Is the Limit: Global Warming as Global Governmentality. *European Journal of International Relations, 19*(1), 69–91.

Naess, A. (1973). The Shallow and the Deep, Longe Range Ecology Movement. A Summary. *Inquiry, 16*, 95–100.

Palme, O. (1972, June 6). Statement by the Prime Minister Olof Palme in the Plenary Meeting. Downloaded June 22, 2022. Arbetarrörelsens arkiv: http://www.olofpalme.org/wp-content/dokument/720606a_fn_miljo.pdf

Säve, P. A. (1915). Sista paret ut! *Sveriges natur, 1*, 1–21.

Scott, T. (1989). *Thomas Müntzer: Theology and Revolution in the German Reformation*. Palgrave Macmillan.

Swyngedouw, E. (2013). Apocalypse Now! Fear and Doomsday Pleasures. *Capitalism Nature Socialism, 24*(1), 9–18.

Thörn, H., Cassegård, C., Soneryd, L., & Wettergren, Å. (2017). Hegemony and Environmentalist Strategy: Global Governance, Movement Mobilization and Climate Justice. In H. Thörn, C. Cassegård, L. Soneryd, & Å. Wettergren

(Eds.), *Climate Action in a Globalizing World: Comparative Perspectives on Environmental Movements in the Global North* (pp. 219–244). Routledge.

Thunberg, G. (2019). *No One is Too Small to Make a Difference*. Penguin.

Yuen, E. (2012). The Politics of Failure Have Failed: The Environmental Movement and Catastrophism. In S. Lilley, D. McNally, E. Yuen, & J. Davis (Eds.), *Catastrophism: The Apocalyptic Politics of Collapse and Rebirth* (pp. 15–43). PM Press.

CHAPTER 4

Postapocalypse

Abstract This chapter presents the postapocalyptic narrative, according to which the apocalypse is already here or inevitable. Emotionally, grief and experiences of loss are central. As a narrative of an unfolding global catastrophe, it came into prominence with the climate issue in the 2000s, but it existed already before that as an undercurrent in many early protest movements. Today, it informs much 'justice' activism and has a strong presence in currents such as the transition movement and the movement for the rights of nature. A fundamental ambivalence in the narrative revolves around what forms of action remain meaningful despite the disasters. While critics have argued that it implies defeatism, we point to the existence of postapocalyptic politics whose action forms range from mourning to mass protest.

Keywords Environmental movement • Narrative • Post-apocalypse • Catastrophe • Transition movement • Dark Mountain • Collapsology

'The misery surpassed our worst fears a hundredfold' (Matsumoto, 2000, p. 18). So the journalist Matsumoto Eiko—writing under the pseudonym 'Midoriko', the green child—described her encounter with the devastation around the Watarase River, an area polluted by sulphur, arsenic, and other toxins from the nearby Ashio Copper Mine. Her series of articles provided the first major report on what is now generally regarded as the first major environmental catastrophe in Japan. Initially published in the daily press in 1901–1902 and then collected in a book entitled *Kōdokuchi no sanjō* (Misery in the Land of Mining Pollution, 1902), the articles played a significant role in garnering support for the growing anti-pollution protests among the farmers of the victimized area, which stretched over three prefectures north of Tokyo. In chapter after chapter, Matsumoto described harrowing encounters with sickly and impoverished families, mothers who had lost their breast milk, and undernourished children. The toxins had also exacted a terrifying toll on nature. A desolate silence had settled over the area that had previously enjoyed a rich fish, insect, and bird life. Vegetation had died along the river including rugged species such as bamboo and willow. The rain that previously nourished the land was a stream of poison, with 'the formerly fertile soil' becoming 'a swamp of mining poisons' (ibid., p. 18).

These passages eerily foreshadow Rachel Carson's famous fable at the beginning of the environmentalist classic *Silent Spring*, which we discussed in the previous chapter. Although Carson points out that everything depicted in the fable had occurred somewhere, her fable above all functions as a warning about future disasters that need to be averted. By contrast, Matsumoto talked little about the future. The devastation she depicted was no fable, but a depiction of a catastrophe that was already reality—of crushed livelihoods and already existing suffering and injustice. These concerns were not merely local. In the words of Tanaka Shōzō—who led the protests among the farmers against the polluting mine and later became known as the 'father' of Japanese environmentalism—the devastation in the Watarase area was the true face of the progress and 'civilization' achieved by the newly rising modern Japan. As the government itself made clear, the Ashio Copper Mine—at the time the largest copper mine in East Asia—was indispensible to Japanese industrial development, not least as a source of sorely needed foreign currency (Strong, 1995, p. 74). By knowingly sacrificing the Watarase area for the sake of industrial development, the government and its allies in the corporate world had, Tanaka repeatedly pointed out, trespassed against the laws of 'Heaven' and thereby indicted the country itself, which had become ruined, a 'dead country'.

The protests against the Ashio Copper Mine not only marked the start of environmental activism in Japan; they were also an early example of postapocalyptic environmentalism. In the latter approach, such catastrophes are not primarily viewed as a future threat but as a reality that must be recognized as a premise for political action. Unlike the narratives of green progress or of the apocalypse, the pivot of this narrative is not *hope* in further economic or technological progress nor *fear* of a future doomsday that must be avoided, but rather grief over losses and injustices suffered in connection with catastrophes that are already ongoing or viewed as unstoppable.

Only in the last decade have postapocalyptic narratives gained a significant following in environmental movements in the Global North and started to be framed in terms of truly global catastrophes, such as climate change. Nevertheless, postapocalyptic experiences in connection with environmental destruction have been articulated throughout the era of industrial capitalism. In Chap. 3, we pointed to a few instances of a postapocalyptic reasoning in both Borgström's and Carson's writings, and many recurring characteristics of the postapocalyptic narrative were clearly visible already among the protesting Watarase farmers.

The first and perhaps most striking characteristic of the postapocalyptic narrative is the central role given to emotions connected to loss such as grief, pain, despair, and anger, as well as practices like mourning. Secondly, the narrative implies a rejection of dominant notions of progress, which is equated with ongoing catastrophe rather than with improvement. To activists in the Ashio struggle, the copper pollution was integral to Japan's embrace of modern civilization, and similarly postapocalyptic environmentalists today view catastrophe as ingrained in the presently dominant capitalist system, which cannot subsist without wrecking the world. That things go on as usual *is* the catastrophe, as Walter Benjamin (1977a, p. 246) wrote.

Thirdly, the narrative fuels what we call a postapocalyptic politics. Recognizing catastrophe as an inescapable point of departure does not imply that 'nothing can be done'. Rather than resulting in passivity or defeatism, such recognition may in fact fuel political action. Indeed, we suggest that most conspicuous attempts by grassroots movements in recent years to challenge institutionalized or established forms of environmentalism give a central place to losses that are already in progress or seen as inevitable. This is due not least to the insistence on justice and morality that often plays a highly activating role in this narrative. Postapocalyptic environmentalism is not so much about giving up action as finding forms of action that remain meaningful despite the experience of catastrophic loss. Such experiences can put the spotlight on conflict lines in society that were previously ignored or less visible. By working through loss, new visions may emerge that can guide activism and provide blueprints for social change. Postapocalyptic environmentalism thus furnishes a model for activism in the *midst* of catastrophe—where the goal is not so much to avert catastrophe as to lessen suffering, restoring and salvaging what is possible, and ensuring justice for those who have been wronged.

Below, we start by tracing the historical background of the postapocalyptic narrative in the Global North and South. We then turn to three examples of this narrative in contemporary environmentalism to explore what nature interests are expressed in them and what kinds of politics springs from this narrative. The three examples are: the struggle for justice as exemplified by the International Tribunal for the Rights of Nature; practices of mourning and preparing for collapse as exemplified by the Dark Mountain collective and similar groups; and the recent wave of climate protests. In our choice of examples, we have chosen to focus on comparatively radical cases of environmental activism. It is true that even institutionalized actors are registering the increasing prevalence of catastrophes—for instance, in the UN-led climate negotiations, the pressure of parties and organizations from the Global South has resulted in adaptation and loss-and-damage being given increasing weight at the expense of mitigation. Still, it is radical environmental activism that offers the clearest examples of the postapocalyptic narrative. In particular, such activism is instructive since it shows how this narrative can be politicized and articulated as a challenge to the status quo.

From the Environmentalism of the Poor to Post-Copenhagen Grief

The postapocalyptic narrative became prominent within the environmental movement in the Global North in the twenty-first-century, in connection with the mobilization against climate change, the increasing concerns about peak oil and energy, and the ongoing mass extinction of species. It is thus easy to view it as a latecomer compared to the narrative of green progress that was established a century ago and the apocalyptic narrative that surged in the post-war period. Yet, the postapocalyptic narrative has deep historical roots and antecedents. As the Ashio struggle shows, it existed already in the early environmental movement, especially among local farmers or fishermen directly victimized by loss of livelihood or health problems caused by pollution. The Ashio struggle has been called an early example of 'the environmentalism of the poor' (Martinez-Alier, 2003, p. 53), and indeed the narrative has often had an affinity with such environmentalism. Protests often appealed to notions of justice informed by traditional value systems and the link to a social nature interest was often clear. Such environmentalism was an important source of the environmental justice movement that sprang up among poor and African American communities in the US in the 1980s, and later of the climate justice movement (Schlosberg & Collins, 2014). Unlike the environmentalist narratives associated with green progress or the apocalypse, justice-related environmentalism has focused on the suffering inflicted on the poor and other disadvantaged groups. In particular, environmental movements in the Global South and among indigenous peoples have often had an orientation to ongoing catastrophes and it is easy to see how their rhetoric anticipates a postapocalyptic narrative.

We can illustrate the above claim with the example of Sámi activism. With their mobilization against hydropower from the late 1950s onward, a postapocalyptic narrative emerged for the first time in the Swedish environmental movement. This happened in the 1960s, when the battle against hydropower appeared to be lost in the eyes of many activists. Grief and a strong awareness of the ongoing destruction of the conditions for traditional livelihoods and lifestyles gained a visible presence in the activism. We find this expressed in a statement by Israel Ruong, chair of the Sámi National Association, who in 1967 described the threat against Sámi culture in terms of 'the industrial age, which hurls itself like a blizzard over the world' (1967, p. 214). The postapocalyptic narrative was also expressed poetically by Inger and Paulus Utsi, who were repeatedly forced to relocate in connection with the construction of water reservoirs in order to harness the hydropower of the Lule River. Their poem 'Shoreless shore' mourns the destruction of a physical habitat as a result of the intrusion of strangers and their boundless needs.

> I understand nothing
> My soul is downcast
> Look around
> The whole village has vanished
> Drowned by strangers
> Their needs have no bounds

I stand on a shore
On a shoreless shore
Look around
Old shores no longer exist
Strangers stole it from us
Their needs have no bounds

Now my people are rocking
On a stormy sea
Look around
Our right is shorn of the protection of justice
Strangers demanded it of us
Their needs have no bounds.
(Utsi & Utsi, 1980, p. 40, our translation)

Yet mourning did not result in passivity. From the 1960s, the Sámi articulated this struggle in anti-colonial terms, stressing parallels and connections with the experiences of other indigenous peoples and anti-colonial struggles. A sense of global solidarity took form, but not in the shape of a homogeneous global 'we', but as a network of solidarity between affected groups—as when the Sámi made connections with the North American indigenous people. Here we can see how the anger over ongoing local disasters gradually took on global significance, with local disasters coming to be seen as instances of global injustice.

The fact that the postapocalyptic narrative is often rooted in the environmentalism of the poor, among people who directly experience the impact of environmental degradation on their lives and livelihoods, has ensured that a social nature interest is often prominent in movement currents employing this narrative. The struggles of the Watarase farmers and the Sámi are cases in point, as are the numerous conflicts around extraction, pollution, pipelines, logging and land conflicts today. To contextualize these currents, we should recall their social geography: they are found more in the Global South than in the Global North, in rural areas more so than in cities, and in poor areas rather than in affluent ones. In other words, they tend to spring from the peripheral areas of the capitalist world economy rather than its core. Often these areas are so-called *sacrifice zones*—spaces physically and culturally destroyed for the benefit of 'the broader good' usually taken to mean industrial progress or the demands of capitalist growth (Klein, 2014, pp. 169ff, 310–315). These zones are usually populated by poor or disadvantaged groups that lack political power and have had to bear a disproportionate share of the burden of a country's industrial development. They are places where polluting factories, power plants, mines, and dams have been located with destructive consequences for the local community. An obvious example is the area around the Ashio copper mine, where the authorities frankly declared that the suffering of the local population was a necessary price to be paid for Japan's industrialization (Strong, 1995, p. 74).

The notion of sacrifice zone helps us understand the wave of protests against fossil fuel extraction today. The numerous actions in recent years against fracking, open-pit mines, tar sands oil pipelines, and so on are to a large extent driven not by fear for the future so much as concern over what is being destroyed here and now, in specific locales and in a way that victimizes specific groups. As Naomi Klein (2014, p. 169f) notes, technologies of extraction depend on the creation of clearly identifiable sacrifice zones—'places that, to their extractors, somehow don't count and therefore can be poisoned, drained, or otherwise destroyed, for the supposed greater good of economic progress'. These places are often 'bound up with notions of racial superiority, because in order to have sacrifice zones, you need to have people and cultures who count so little that they are considered deserving of sacrifice'. In Rob Nixon's (2011, p. 17) words, 'the exponential upsurge in indigenous resource rebellions across the globe' in the epoch of neoliberalism largely revolves around:

> [...] a clash of temporal perspectives between the short-termers who arrive (with their official landscape maps) to extract, despoil, and depart and the long-termers who must live inside the ecological aftermath and must therefore weigh wealth differently in time's scales.

As activists testify, the destruction attendant on sacrifice zones can be experienced as the very end of a world, because world is often inseparable from place—from rivers, forests, and lands central to the identities of the victimized groups. This, for instance, was expressed by the Standing Rock Sioux as they stood up as 'water protectors' to protest the planned Dakota Access Pipeline (Streeby, 2018, p. 39f). Ironically, sacrifice zones are today created not only for the sake of fossil fuel extraction, but also for extracting minerals needed for renewables—a fact that highlights how even 'green' progress can be experienced as catastrophic.

While a clear recognition of ongoing catastrophes has always been prominent in the environmentalism of the Global South and among indigenous peoples, in the Global North it was the climate issue in the 2000s that for the first time brought a postapocalyptic narrative to the fore. Concerns about peak oil, the accelerating loss of biodiversity, and the nuclear catastrophe in Fukushima in 2011 also fed into this development. The postapocalyptic environmentalism that emerged in relation to the climate was emphatically framed in global terms, nourished by the realization that it may already be too late to stave off global climate breakdown. Such postapocalyptic views have an objective background, namely the failure of environmentalists, governments, and other actors to stop global warming, despite decades of climate summits, high levels of public environmental awareness, and a scientific consensus about the causes of this degradation. It is thus a fruit of the paradox of the Anthropocene: the fact that a humanity that has seemingly become more powerful than ever, even to the point of changing the course of the earth, is utterly unable to unshackle itself from a system that seems hellbent on self-destruction (Hamilton, 2017, p. viif; Stoner & Melathopoulos, 2015, p. 18).

A pivotal event that spurred this development was the failed UN climate summit in Copenhagen 2009 (COP 15), which resulted in widespread disillusionment and partial demobilization in the environmental movement (Cassegård et al., 2017; de Moor & Wahlström, 2019). Already before this summit activists and critics mobilizing around climate justice had increasingly started to voice criticism of the narrative of green progress and that of the apocalypse that were prevailing in the environmental movement. This criticism must be understood against the backdrop of the partial incorporation of the apocalyptic narrative into the establishment rhetoric of climate change. When heads of government, billionaires, and establishment figures like Al Gore routinely invoked disaster scenarios while advocating market mechanisms and new technology to solve the climate crisis, it became clear that the apocalyptic narrative was not necessarily subversive. By presenting the apocalypse as a mere future threat that could be staved off by the proper application of technology and economic incentives, it conveniently obscured the fact that the apocalypse was *already here*, especially for many in the Global South.

Clearly, growing disillusionment with the institutions of climate governance fed into this criticism. The failed climate negotiations in Copenhagen further dented the credibility of the green progress narrative as well as the establishment version of the apocalyptic narrative in the eyes of many activists. Around the world, groups began to appear that pointed to the reality of ongoing disasters and openly declared that they no longer saw any hope of stopping the global catastrophe—whether this was perceived in terms of global warming or a general societal collapse caused by the end of the fossil fuel economy. This despair was reinforced by the increasing frequency of extreme weather events around the world, such as droughts, hurricanes, and forest fires, as well as by concerns about energy and the ongoing 'sixth' mass extinction.

The sense of despair also resonated with general cultural trends. In film and literature, postapocalyptic scenarios (Berger, 1999) have grown increasingly common and are often noticeably influenced by climate anxiety or so-called 'climate trauma', described by cultural critic Ann Kaplan as 'pre-traumatic stress' induced by 'anticipatory anxiety about the future' (2016, xix; see also Richardson, 2018). Such scenarios are also increasingly recognized as realistic by many people, as seen in the fashion for survivalist 'prepper' communities (Bounds, 2021), 'prepper' merchandise and TV shows (Foster, 2016), as well as academic and non-fiction works dealing with collapses or a post-catastrophic world. Academia saw a proliferation of new coinages—such as solastalgia (Albrecht, 2006), ecological grief (Cunsolo & Ellis, 2018), and planetary hospice (Baker, 2014)—meant to facilitate the verbalization of the spreading feelings of grief and melancholia related to the climate and the environment. Highly visible controversies were also occasioned by essays such as Roy Scranton's *Learning to Die in the Anthropocene* (2015) and Jonathan Franzen's 'What If We Stopped Pretending?' (2019), which both argued that climate apocalypse had become inevitable and urged us to prepare for the inevitable hardships ahead.

Struggling for Justice

One of the most important means to formulate challenges to the status quo within the postapocalyptic narrative has been through the notion of justice. In justice activism a central claim is that *past and present* inequalities and power relations are at the heart of today's severest problems, rather than future threats to the globe as a whole. An important root of this activism has been the struggles of communities in the Global South and other peripheral areas to defend their livelihoods, local autonomy, and customary uses of nature—struggles in which social and moral nature interests have been central.

In such struggles, participants regularly need to address a host of issues and conflicts that are not environmental in a narrow sense. That disasters occur in some places but not others, that some are hit harder than others, and that those who bear the greatest responsibility for disasters are rarely among the hardest hit almost inevitably means that power and dominance become focal points of concern. For instance, the struggles of the Sámi and other indigenous peoples link the environmental crisis to land rights, historical injustices, and the defence of traditional life forms. A recent and spectacular example of the broad range of social issues that can be actualized by a disaster is the protest wave following the 2011 nuclear meltdown in Fukushima (Cassegård, 2014, pp. 244–248). The region around the plant was in many respects a typical sacrifice zone, a rural periphery in relation to the metropolitan economy of Tokyo which it supplied with electricity. Economically backward, Fukushima had accepted the plan to host the nuclear plant in the 1960s due to the prospect of economic revitalization and the financial compensation promised by the central government. Lives, livelihoods, and health were directly affected by the disaster, which led to a brutal collapse of farms, industries, and local communities in the vicinity of the power plant and triggered a wave of 'nuclear refugees'. As participants in the protests put it, the movement concerned far more than the environment and energy in a narrow sense; it was also a broader 'democracy movement' protesting the exercise of power in Japan, which meant that issues related to discrimination, working conditions, and rural neglect became central points of conflict. A variety of social ills were thus highlighted by these protests, testifying to the centrality of the social nature interest. To the extent that protesters expressed hope, it was a hope for justice—that power structures would be abolished, that the culpable would be punished, and that victims would receive the help they needed. The needs of victims were prioritized over the pragmatic adaptation to what the system could deliver, meaning that the demands of the protesters (such as 'Give back Fukushima', or in other words: give back what is irretrievably lost) polemically turned into an accusation of the system itself. While demands were made for economic compensation and support to communities, protesters were also adamant that such measures would be insufficient since *nothing* could compensate for the losses. The demand for justice was important both for reasons of livelihood and for symbolic and moral reasons: as long as the government and energy companies refused to accept their responsibilities, the wounds would never heal, nor could the broken social contract be restored.

One of the clearest and most explicit attempts in recent years to articulate justice in relation to ongoing environmental disasters is the campaign for the Rights of Nature, which is rooted in the environmentalism of the poor and the anti-colonial struggle of indigenous peoples. These rights are famously enshrined in the constitutions of Bolivia and Ecuador, and they also underpin the ongoing efforts to establish a legal framework tied to 'ecocide', a term coined in reference to US warfare in Vietnam at the 1970 Conference on War and National Responsibility in Washington. Here a juridical nature interest supplements the social one, since nature is explicitly viewed as a bearer of rights. Indigenous peoples have often sought, and received, support from the UN in their struggle against colonial nation-states, and the juridical nature interest can be seen as a means of invoking the Declaration of Human Rights while demanding its extension, not only to living beings, as in the animal rights movement (which has also been the subject of intensified transnational mobilization since the 1990s), but also to rivers and forests.

Participants in this campaign do not merely call for the extension of existing legal frameworks to include nature. They also invoke them polemically to highlight the insufficiency of these frameworks. A good example is the usage of the rights of nature in the International Tribunal for the Rights of Nature, which is closely linked to the struggles of indigenous peoples for whom the attempt to defend their autonomy and the rights of Mother Nature are a prolongation of their resistance against colonialism. The tribunal, which held its first session in Quito in 2014, is a mock tribunal set up by activists. While lacking legal force, it adopts the trappings of regular courts, including prosecutors, expert witnesses, a panel of judges, the presentation of evidence, and clearly defined cases. Solemn rituals may also be performed with roots in indigenous traditions to put participants and onlookers in the right frame of mind. Following the Quito tribunal, sessions have so far also been held in connection with the UN climate summits in Lima, Paris, Bonn, and Glasgow, in activist spaces outside the summit venues, in addition to a number of regional sessions.[1]

Setting up the tribunals has been an important tool for activists both as a news-grabbing spectacle and as a convenient platform for voicing different grievances and telling different stories related to climate change and

---

[1] Our analysis in this section builds on fieldwork during the tribunals in Lima 2014 and Paris 2015 (see Cassegård & Thörn, 2018). Background on the ITNR can be found on: http://therightsofnature.org/

climate governance, presenting them as part of a unitary movement. In the proceedings, cases are treated that involve polluted water, deforestation, oil spills, hydroelectric dams that will flood communities and cause major displacement, murdered anti-mining activists, 'man-made earthquakes' caused by fracking, the destruction of the Great Barrier Reef, the oil extraction in Yasuni in Ecuador, and UN-sponsored attempts to commodify nature such as REDD+. Unifying the accusations brought forth before the tribunal is that they are animated by outrage at losses that have happened in specific locales in ways that victimize specific groups. Almost all cases involve historical injustices and rights in relation to sacrifice zones. Grievances experienced at specific localities often take precedence over concern for the globe as a whole. A message repeated in various forms is that suffering is taking place, not in the future, but right now—a message condensed in the image of an agonizing Mother Earth, or Pachamama. Mother Earth, the judge Osprey Orielle Lake stated during the Paris tribunal in 2015, 'is calling out to us and our hearts are breaking' (quoted in Cassegård & Thörn, 2018, p. 572). Rather than preventing future threats, the struggle is to halt the suffering and seek justice for crimes already perpetrated so that a process of healing can start. Regarding temporality, there is thus a clear break with the future-oriented perspective characteristic of mainstream apocalyptic environmentalism as well as of the future-oriented optimism seen in the green progress narrative.

The tribunals' close link to the struggles of communities in the Global South seeking to defend their livelihoods, autonomy, and customary use of nature make them an integral part of the broader climate justice movement. The collective 'we' takes the form of a translocal solidarity of victimized communities—usually in the Global South, among indigenous people and among the rural poor—that unite with nature against the perpetrators, which include global corporations, governments, and, in part, the UN. The ability of the tribunals to bring up cases from anywhere on the globe make them well suited to visualizing this broad, translocal solidarity. Yet while local struggles converge in translocal solidarities, the local remains pivotal as a touchstone of the rightness of the struggle, since it is on the local level, in the sacrifice zone, that the decisive loss or violation is inflicted in response to which justice is sought.

More emphatically than many other groups in the climate justice movement, the tribunals stress not only a social nature interest but also an ecological nature interest in the sense that the justice they fight for is not just that of people, but of all living things and the earth itself. The fundamental idea behind the tribunals—that nature is a subject of rights—is thus biocentric. Their goal is to promote and protect these rights, as well as those of communities that are part of, and live in harmony with, the earth. The sharp boundary between the human and the non-human is thereby erased. The rights of nature constitute part of a 'bio-humanistic practice that recognizes nature as both material and existential'. Asserting such rights is not a call to 'conserve' wild flora and fauna, since they include the rights of communities living with nature as 'part of the vital and interdependent movement of nature' and thus as part of the same ecosystem (quoted from the tribunal pamphlet, Paris 2015). Rather than first nature, it is a second nature shaped through a long history of interdependence between communities and their environment that is made the subject of rights. The rights thus cannot be used to damage the life of communities with the excuse of protecting nature. The proceedings are often directed precisely against policies and initiatives that, while ostensibly protecting the climate or nature, prevent local communities from exercising traditional lifestyles or that directly destroy their possibilities of maintaining traditional livelihoods.

Let us take a closer look at how the justice that the tribunals invoke relates to the justice embodied in existing law. This will give us a clearer understanding of their confrontational nature. 'This is a space that replaces the governments', as Natalia Greene, an activist in the Global Alliance for the Rights of Nature, stated in a speech after the Lima tribunal (quoted in Cassegård & Thörn, 2018, p. 572). The message of participants is not that a tribunal should be established, but that it *has* been established. Like all mock tribunals, the tribunal offers an alternative to the existing legal system, which is perceived to have failed to uphold justice. It thus impersonates the missing institution of justice itself, delivering a verdict not only on the various cases brought before it but simultaneously on the existing system of right.

The ambiguity of this position is that it both mimics and transcends the existing judicial system. The notion of justice it brings into play approaches that of Jacques Derrida. Building on Walter Benjamin's distinction between right and justice, Derrida (1990, p. 947) claims that justice exists as an excess in relation to the formal system of institutionalized law. There is always a tension—or 'justice gap' (Balibar, 2012)—between the emergent and not-yet-captured qualities of justice and the formal systems of law, which can be brought into play by activists to destabilize the latter. Justice is a conflictual concept that becomes visible in moments of struggle when those who are excluded by law claim justice. The tribunals thus represent a way for activists to hold on to the idea of justice even while standing outside and condemning the existing legal system. Responding to experiences of destruction and loss which existing institutions have been unable to prevent and in many cases have even aggravated, the justice of the tribunals takes on apocalyptic qualities in the sense that it reveals the untruth of the existing institutions of human law, of governments, of capitalism, and of the international climate governance regime. The utopia implicit in the tribunals is that of the longed-for arrival of justice, which heals, restores, and punishes but which is impossible within the confines of the existing legal system. This rejection of existing institutions is reinforced when, as in the case of the tribunals, it draws on centuries of colonization and oppression. Justice, in the sense in which it is employed by Derrida, becomes a way for activists to express outrage at the injustices that they see no prospects of ever being dealt with adequately by existing systems. It is needed to express the excess that goes beyond monetary compensation—the grievances of humans and animals who have already died or had their health irremediably damaged, and nature that has been irreparably destroyed, to wrongs that cannot be righted by any human court. 'A wound on mother earth cannot be compensated. It must be healed', as one of the judges, Tom Goldtooth, declared during his concluding speech in Paris in 2015 (quoted in Cassegård & Thörn, 2018, p. 573). In this sense, as Derrida (2006, p. xviii) points out, 'justice' is linked to ghosts, to those who have a rightful claim to justice which it is *too late* for them to receive, since the wrongs are irreparable.

An important point of the tribunals is to give voice to such ghosts, to have a subversive language recognized and listened to—a language which the dominant legal institutions are incapable of hearing. As Balibar (2012, p. 27) notes, claiming justice requires the excluded to speak—hence the problem that the wrong is 'redoubled by the fact that it cannot become expressed in the dominant language of the judge'. As long as the subject is captured in existing institutions, it is forced into a subaltern condition in which it is deprived of autonomous means of expression. The tribunals are, in this respect, meant to function as a necessary corrective, an arena in which justice claims can be formulated, rehearsed, and amplified, and, from there, circulated to the wider public, where they may also be heard by and put pressure on those who are responsible. In its polemical and accusatory gesture, this iteration of postapocalyptic politics offers a contrast to the somewhat depoliticized cultural activism to which we will now turn.

MOURNING AND PREPARING FOR COLLAPSE

We now turn to another current of postapocalyptic environmentalism, more prominent in the Global North. Rather than struggling for justice, here the stress is on culturally processing the realization that today's civilization cannot be sustained and on preparing for the collapse. In terms of content, this activism has aimed at spreading awareness about the likely demise of civilization as we know it, challenging the ideology of progress, creating art and poetry that may guide us in an unknown and more 'uncivilized' world, and practically developing lifestyles that make us less dependent on the complex systems of modern society. An ecological nature interest has often been evident in this iteration of activism (although aesthetic and moral ones also play a role, as we will see). This points to a continuity with the apocalyptic narrative, but unlike the latter, this approach presents the collapse as too late to stave off. Activism must therefore revolve around adapting to the impending breakdown with a minimum of suffering, for both people and other forms of life.

The Dark Mountain collective in the UK was one of the first groups to exemplify this current. It was created in 2009 as a network or writers, artists, craftspeople, and others and has since been engaged in arranging meetings, workshops, and festivals, as well as publishing a manifesto and a series of anthologies with essays, short stories, poems, and pictures. It is explicit in accepting the collapse of our civilization as inevitable. 'We believe we are entering an age of material decline, ecological collapse and social and political uncertainty, and that our cultural responses should reflect this, rather than denying it' (Dark Mountain Project, n.d.). The stance is eloquently expressed by the group's two founders, Paul Kingsnorth and Dougald Hine, in their 2009 'Uncivilization' manifesto. They explain that what motivated them to start the group was disappointment in established environmentalism—the fact that none of its campaigns were succeeding and that environmentalists were not being honest with themselves. Instead of fearing the coming climate change with its attendant decline, depletion, chaos, and hardships, they embrace it:

> Together we are able to say it loud and clear: we are not going to 'save the planet'. The planet is not ours to save. The planet is not dying; but our civilisation might be, and neither green technology nor ethical shopping is going to prevent a serious crash. Curiously enough, accepting this reality brings about not despair, as some have suggested, but a great sense of hope.

Once we stop pretending that the impossible can happen, we are released to think seriously about the future. (Kingsnorth, 2010)

Dark Mountain's main activity is cultural—providing new words, poems, stories, and artistic expressions equal to the losses we are facing that may aid us in the necessary mental reorientation away from the 'myth of progress' in general and the narrative of green progress in particular (Kingsnorth & Hine, 2017, p. 17, 23; Du Cann et al., 2017, p. 1). To participants, this expressive work is a work of mourning—a process involving the verbalization of loss. Kingsnorth (2017, p. 98) writes that Dark Mountain offers a 'planetary grieving process', and that to him the group became 'a way to work through the grief caused by the end of so much of what we hold dear'. Hine (2017, p. 244) describes the group's activities as a process of recovering from 'cultural bereavement'.

The collective 'we' shaped in the group's activities appears to be formed firstly by an ecological nature interest, involving an ecocentric identification with and despairing love for the vanishing nature and the wild, which in the group's manifesto is expressed as the desire to escape the 'human bubble' and fashion a perspective on humanity which other species—'a blue whale, an albatross, a mountain hare'—might recognize as approaching the truth (Kingsnorth & Hine, 2017, p. 274). This is a universalism wide enough to include nature along with those forms of human togetherness that might survive civilizational collapse and provide space for hope. Secondly, the group's 'we' is shaped by the rejection of 'neo-environmentalism', described as a neoliberal environmentalism friendly to markets and green technology legitimized through the jargon of 'sustainability' (see e.g. Kingsnorth, 2017, pp. 61–82, 131ff). In explicit opposition to that tendency, Kingsnorth announces: 'I withdraw from the campaigning and the marching' (ibid., p. 81). This stance has earned Dark Mountain criticism from other environmentalists for being fatalist or defeatist. The ensuing debates saw Kingsnorth explicitly reject the upbeat future-oriented optimism of 'professionally optimistic' environmentalists who 'believe that these things should not be said, even if true, because saying them will deprive people of "hope", and without hope there will be no chance of "saving the planet". But false hope is worse than no hope at all' (Kingsnorth & Monbiot, 2009).

Dark Mountain shares the ecological nature interest with large parts of the established environmental movement in the Global North. At the same time, the group combines this nature interest with others that demonstrate a historical continuity with the early nature conservation movement. In the desire to escape the 'human bubble', we recognize an aesthetic and a moral nature interest—aesthetic since the wilderness still provides the standard of beauty, and moral since the approaching collapse will force us to reflect morally and existentially on ourselves and on what kind of people we want to be when things fall apart. In this confrontation with an unknown darkness beyond our civilization, group members like Hine detect a utopian spark—a hope, even, for a more genuine self. Speaking of the encounter with 'the dark knowledge of climate change' as a liminal experience that makes all our previous knowledge unreliable, he continues:

> And so the question I've been sitting with is: how is the space to be held, what can we learn from, who can we learn from, where are the stories and the practices that we might need in order that this can be an initiation, a transformation that has some hope of finding paths onwards into the unknown world that lies ahead, beyond the unravelling of the world as we knew it? (Hine, 2020)

Hine's own answer to this is simple: 'If someone were to ask me what kind of cause is sufficient to live for in dark times, the best answer I could give would be: to take responsibility for the survival of something that matters deeply' (ibid., 2012).

The outsider perspective on human civilization advocated by Dark Mountain is reminiscent of Ricoeur's description of Utopia that was discussed in Chap. 1, as a 'no-place' from which an exterior glance can be cast on ourselves and our society, which in its light is revealed as strange and possible to criticize. Adorno (1978, p. 247) described this as an attempt to fashion perspectives 'that displace and estrange the world, reveal it to be, with its rifts and crevices, as indigent and distorted as it will appear one day in the messianic light'. In Dark Mountain's case, the exterior glance permits a distancing from the 'bubble' while highlighting the devastating consequences of civilization for humans as well as other life forms. In elegiacal passages, Kingsnorth (2017) laments the demise of a beloved wilderness—hawks circling the sky, sunsets, and untamed rivers. In this way, a romantic concept of nature is embraced that is distinct from that of the justice movement, which has usually oriented itself towards second nature, that is, a nature shaped by interaction with humanity.

While Dark Mountain practitioners may have withdrawn from conventional environmentalism, they have re-engaged in other struggles. Kingsnorth (2017, p. 118) states that rather than being about defeat, withdrawal is about 'pulling back to a place where you can find the breathing space to be free and human again'. His withdrawal, he claims, has made him more engaged than before: 'As far as I'm concerned, I'm doing more than I've ever done' (quoted in Wilt, 2014). Much of his time is spent managing his two-and-a-half acres of land ecologically and he also offers courses on how to use a scythe. He compares his attitude to that of a poppy, whose seed can lie dormant for 80 years, even if paved over, built on or oversown:

> Be a poppy then, in the face of the Machine? It seems a good task to set myself. [...] In an age of loss, our task is surely to keep safe what we can when the Machine passes by, hungry and howling for blood. To be still and stoical and protective, to pass on truths and skills that will always be truths and skills, and never forget to remember what we are losing, every day that we live. (Kingsnorth, 2017, p. 89)

In pulling back to the local to pursue practical work, Kingsnorth follows in the steps of many participants in other groups and networks similar to Dark Mountain that have emerged in the last decade. One of the best known is the collapsology network, which originated in France in 2013 and stresses the need to prepare for societal collapse mentally and practically (Tasset, 2019). Collapsology, according to the well-known spokesperson Pablo Servigne and his colleagues, is not only the study of collapse but also a way to emotionally process it, search for new relationships to nature, and act to minimize suffering and injustice. According to the group, the collapse has already started and is ongoing, although not geographically uniform. It will be an unpredictable process, with domino effects, arising from simultaneous interlocking crises (Servigne & Stevens, 2020; Servigne et al., 2021). Mindful of the criticism directed against Dark Mountain and similar groups for defeatism and passivity, they stress the need for action and that 'we have to fight' even if the future is dark:

> To be catastrophist, for us, simply means avoiding a posture of denial and taking note of the catastrophes *that are taking place*. We have to learn to see them, to accept their existence and to say goodbye to all that these events will deprive us of. In our opinion, it's this attitude of courage, of conscious-

ness and calm, with our eyes wide open, that will enable us to pursue realistic future paths. This isn't pessimism! (Servigne & Stevens, 2020, p. 177f)

Another influential network is the Deep Adaptation network initiated by Jem Bendell, a professor of sustainability, which has spread to numerous countries and influenced groups such as Extinction Rebellion (see Bendell & Read, 2021). Deep adaptation refers to the personal and collective changes that will be necessary to adapt to the near-term social collapse that is likely or inevitable due to climate change. In view of this collapse, Bendell and other participants in this network stress the importance of grief, love, learning, and compassion. More than in Dark Mountain, there are attempts in both the collapsology and Deep Adaptation networks to reach out to and find ways of collaborating with movements such as the climate justice movement and the transition movement.

The transition movement too is strongly coloured by postapocalyptic environmentalism, although it differs from the above-mentioned groups through its stress on positive feelings and its much heavier emphasis on local, practical work. It harkens back to the first 'transition town' in Totnes in the UK in 2006 and has since spread around the world. Its chief aim is to increase resilience on a local level by preparing practically for the hardships that will be inevitable in the future due to global warming and resource scarcity. It also stresses the necessity of what it calls an 'inner transition' to adapt mentally and emotionally to the changes, to combat feelings of anxiety, sadness, and powerlessness, and to let go of ideas of endless growth and progress—a mental transition that is often accompanied by practices such as yoga and meditation. Although having little hope in the present capitalist system, which is unsustainable and bound to collapse, transitioners generally avoid overt criticism, preferring constructive work to build a better alternative. As participants put it, they want to be a movement *for* rather than a movement *against*. More than Dark Mountain, they tend to stress the brighter or more hopeful aspects of the collapse, emphasizing the historical opportunity to create a better society and achieve degrowth without social chaos. Hope is generated by acting together and in contact with nature, often in practical projects around issues such as energy supply, food production, and skills sharing. The transition movement is thus able to take key elements of a postapocalyptic narrative—the inevitability of climate change and resource depletion and the abandonment of the idea of growth/progress—and reframe them in a way that makes it possible to experience them positively.

With its emphasis on locally based practical activities in close contact with nature as well as with other people, the transition movement illustrates the central place that a moral nature interest can acquire in postapocalyptic narratives. Rather than wilderness, it is clearly oriented towards second nature, a nature that is the object of work and a source of livelihood. Such nature is also a source of moral value and spiritual wellbeing. Being able to produce one's own food or other necessities removes some of the anxiety about living in an insecure world while nourishing a practically grounded understanding of nature's limits. Interacting with nature opens up an arena where action may make sense despite the disasters, not least since people here feel more directly and palpably than in mainstream society that they are in a world in which they and their surroundings mutually respond and adapt to each other, or that they 'resonate', to use Hartmut Rosa's (2019) term. Daily chores that require constant contact with nature and other people become a way to restore the inner balance, manage emotions, and regain hope. This is not a hope of being able to avert disasters, but rather a vague hopefulness rooted in activity itself for the alternatives to the present system that might sprout through interacting with other people and with nature. Far more than participants in Dark Mountain, for instance, transitioners build on the insight that activity itself can be a way of regaining hope, where more passive or purely intellectual approaches easily lead to despair or practical paralysis.

CALLING FOR EMERGENCY

Unlike the comparatively non-confrontational activities of groups such as Dark Mountain or the transition movement, the wave of climate protest represented by groups such as Fridays for Future and XR is explicitly confrontational and politicizing, in the sense of publicly dramatizing a conflict to force drastic and radical social change. This also means that for them, the expression of anger is just as important as that of mourning. Their system-critical intent is clear from the fact that they insist that the 'emergency brake' must be pulled, even at the cost of disrupting the prevailing system. As expressed by Greta Thunberg, realism in the sense of political pragmatism is woefully insufficient when nature is considered: 'Until you start focusing on what needs to be done rather than what is politically possible, there's no hope' (2019, p. 16). Similarly, Rupert Read, an activist in XR, emphasizes that the demands of the group are 'impossible' demands that cannot be reconciled with business as usual. 'They could only be accommodated by putting in process a revolutionary transformation in our entire way of life' (Read & Alexander, 2019, p. 27).

As Ben Anderson (2017) points out, the very act of declaring an emergency presupposes a minimum of hope that there is still room to act. In that sense, significant elements of the apocalyptic narrative still exist in both XR and Fridays for Future (see e.g. de Moor, 2021, p. 11f; de Moor et al., 2020, p. 5). Still, we believe that it makes sense to view XR as largely postapocalyptic, while Fridays for Future has oscillated between an apocalyptic and a postapocalyptic narrative. XR's articulation of a postapocalyptic narrative is arguably visible in its name, which signifies a rebellion against an ongoing disaster that strikes not only against non-human animals but also against humanity itself, since the extinction of species is already fundamentally undermining the planet's ecological balance, and thus the conditions on which humans depend. XR's preference for the colour black (rather than green) in its public manifestations can also be counted as a postapocalyptic element, as can its so-called 'regenerative culture', with its emphasis on an ethics of care and its use of sessions and workshops to channel grief and despair into action (Westwell & Bunting, 2020). As a direct action-oriented movement, XR has also staged street performances that visualize the ongoing catastrophe and publicly express feelings of grief and anger. For example, activists have on several occasions placed themselves under a gallows pole, standing on a block of ice with a snare around their necks; the snare slowly tightening as the ice melts. On Good Friday in 2019, XR activists in Gothenburg appropriated the Christian feast of mourning by organizing a 'March of Mourning for the Mass Extinction'. Dressed in black clothing they carried a coffin through the streets of the central city, eventually stopping in front of the City Council, where a funeral ritual was held. A prayer was written for the occasion, but Christianity's inward-looking emphasis on submission, guilt, repentance, and dreams of a Godly Kingdom were replaced by strongly expressed emotions of mourning, anger, and demands for truth and political action:

> We mourn the caribou, the white sork, and the black
> We mourn that we have deprived our children of the possibility to have children
> We mourn the ices of Himalaya and the steaming Amazonas
> We mourn those who die in climate catastrophes without even having had the taste of abundance
> We accuse the legislators for their cowardliness

WE DEMAND TRUTH, COURAGE AND POLITICAL ACTION
(our translation)

On the cover of the prayer booklet that was handed out by the activists, the shapes of a male and a female body, signified as 'Homo Sapiens' was visible under a headline that stated 'R.I.P.' (Rest In Peace), followed by a Christian cross and a question mark.

While we cited apocalyptic language used by Greta Thunberg (2019) in Chap. 3, there are also many examples of how she links up with a postapocalyptic narrative in her speeches. This is perhaps most clear when she (repeatedly) uses the expression 'our house' as a metaphor for the planet—especially in her oft-cited exclamation 'our house is on fire'. This expression was used when she spoke on Parliament Square in London on 31 October 2018, when XR announced itself publicly for the first time by presenting its Declaration of Rebellion. On this occasion, Thunberg claimed that what made her environmentally conscious was the sense of unreality that hit her as she realized that on the one hand we are living in an ongoing disaster, in which 200 species are extinguished every day, and on the other hand this disaster is completely denied not only by the media but also by the politicians who have a particular responsibility to act in order to stop it (ibid., p. xx). Further, when Thunberg states that she does not want politicians to feel hope but panic, she expresses the sensation that the catastrophe is already here. The same sensation underlies the call for drastic emergency action that is their central demand and it is nourished by a realization that the catastrophe is here.

HOPE AND MORALITY

As we have seen, the postapocalyptic narrative is far from uniform. A clear dividing line exists, for example, between how it has been articulated in the Global South and North. All the versions of the narrative nevertheless share their point of departure, namely grief at ongoing or inevitable catastrophic destruction. A fundamental ambivalence in the narrative concerns what forms of action remain meaningful despite the disasters and what room there still is for such action. This ambivalence can be expressed through a tension between grief and the need for mourning on the one hand and action-oriented emotions such as anger and hope on the other. Above we saw how important anger is in fuelling the often confrontative direct action approach of XR.

To illustrate how activists deal with this ambivalence, we will now turn to the role of hope, the emotion that more than any other is associated with the utopian action forms that define social movements. Research into the emotions that motivate people to engage in social movements often stresses that hope in some form is necessary for activism (Aminzade & McAdam, 2001, p. 31; Gamson & Meyer, 1996), but the issue is complex. Ghassan Hage (2003) observes that hope is unevenly distributed in capitalist societies, and this uneven distribution also marks the environmental movement. Environmental activists in the Global North have often had access to hope in ways that activists in the Global South have not. Instead of hope, the latter have often relied more on other emotions such as anger, pain, and grief (Kleres & Wettergren, 2017). This shows, firstly, that hope is clearly not always the primary motivating emotion behind activism. At the same time, the growth of postapocalyptic environmentalism in the Global North indicates that many environmentalists in the North too are now turning away from hope, at least in terms of the specific sense that catastrophe can be averted. But if hope can no longer mean hope in averting catastrophe, what form does it take?

The currents associated with a postapocalyptic narrative have often diverged precisely in how they answer this question. The role of hope varies depending on which current of postapocalyptic environmentalism one considers. Below, we will pay particular attention on the one hand to how *new forms of hope can be generated through actions that at first sight might not appear political,* and on the other hand to explicitly *political action that rests on other motivations than hope.* These two contrasting forms of relating hope and activism coexist in today's postapocalyptic environmentalism and give rise to quite different narratives. We start by discussing the vague, non-representational hopefulness that one finds in groups such as Dark Mountain and the transition movement, where it is closely tied to mourning and practical work. We then turn to more explicitly political forms of action and show that these are often motivated by moral reasons that do not necessarily rest on hope.

The Paradox of Hope

In Dark Mountain, we have already encountered one way in which mourning can play a role in activism by facilitating the *recovery* of hope. Dark Mountain presents the acceptance of loss as a positive achievement that is liberating as well as enabling. As Kingsnorth stated in a debate:

> This may sound a strange thing to say, but one of the great achievements for me of the Dark Mountain Project has been to give people permission to give up hope. [...] I don't think we need hope. I think we need imagination. (Stephenson & Kingsnorth, 2012)

Imagination can help people move out of despair, paralysis or 'melancholia', and thereby aid recovery. As one Dark Mountain activist put it: 'Once false hopes are shrugged off, once we stare into the darkness that surrounds us, we can see new paths opening up; new forms; new words' (Du Cann et al., 2017, p. 2). The outcome of this is described as a new form of hope that, as Hine (n.d.) writes, 'may yet lie on the far side of despair'. These passages illuminate what we call the paradox of hope—the fact that hope can sometimes only be gained denying hope, or ridding oneself of 'false' hopes.

However, Dark Mountain is hardly clear about the content of this hope 'on the far side of despair'. Rather than an optimistic assessment of possibilities, it appears to consist in an indistinct sense of hopefulness, nourished by imagination. Using a term coined by the philosopher Jonathan Lear (2006), we could call it 'radical hope', i.e. a form of hope that remains possible even in the face of catastrophes that seemingly destroy all valid grounds for hope, but which must be inarticulate since the cultural resources for endowing it with meaning have been shattered. By being non-representational, such hope is compatible with bleak intellectual assessments about the unavoidability of collapse.

If this hope is a precarious achievement for Dark Mountain participants, it has a stronger and more stable presence in the transition movement. Here hope seems to be generated and sustained by activity itself, through practical work and togetherness. More concretely, it is created practically, by doing or acting, rather than intellectually. The possibility of such active hope is also stressed by collapsologists: action can change our imagination and help us break out of helplessness better than merely intellectual efforts; action is therefore *part* of the process whereby hope is regained, rather than its goal or culmination (Servigne & Stevens, 2020, p. 167). One source of hope often mentioned is nature as encountered through the senses in local, practical work—in being surrounded by nature, in being in contact with living things, or through interaction with different materials. Another source of hope is togetherness, as expressed in activities such as working together or sharing food or other items. Regarding the larger world, one interviewee explains that she engages in practical work to 'feel better' but has little hope for the world. Still, she sees the practical activities as exerting a small, but good influence, which she compares to ripples spreading on a water surface (Swedish transition activist, interviewed 12 October 2020). In the transition movement, this hopefulness is primarily channelled into practical constructive work on a local level and goes hand in hand with relinquishing overt political action in the sense of protests or other forms of public confrontation. Yet, as we shall see in the next section, another way in which activists who have lost hope and withdrawn from politics can develop is by returning to politics and finding ways to engage in it that do not rest on hope.

Morality and Dignity

Many postapocalyptic environmentalists engage in political action based on other motivations than hope. To understand this, we must recall that hope is not the only emotion that can motivate activism. In interviewing collapsologists and transitioners, we learn that anger and sometimes also sorrow, love, and compassion may figure prominently as motivating emotions. Especially striking is the prominence of *ethical* reasons for acting when hope for avoiding disaster is slim or non-existent. Morality, we suggest, may well be the most important answer to why action can be experienced as meaningful despite catastrophic loss. Not only is morality central in justice struggles, but groups such as XR have also been described as being driven by a 'virtue ethics', where activism rests less on optimism regarding the outcome than on the desire to act morally and to do 'the right thing […] regardless of the outcome' (Stuart, 2020, p. 498). A similar stress on virtue can also often be found among collapsologists and transitioners, when they frame their activities in political terms. A moral reflexivity stimulated by the awareness of ongoing or inevitable catastrophes is clearly a common denominator for the more politicized currents of postapocalyptic environmentalism and testifies to the powerful influence of a moral nature interest. One interviewee, a transitioner, thus explains that guilt and a sense of duty towards his children made him try to be 'the person the planet needs'.

> The relevant question now is how to face the collapse: How do we want to be when we know this about the world? How do we want to be? How do we want to face it? How do we want to encounter our future? (Swedish transition activist, interviewed 9 February 2021)

Another activist bluntly admits that he does not necessarily fight to win. More important is the self-respect and 'dignity' that fighting alone can guarantee:

> We are saying that we struggle for two reasons. In part, it is to win in the matter at hand. But it is also to keep our dignity. [...] And that struggle we always win—the struggle for dignity, for human worth, for one's own worth. (Swedish transition activist, interviewed 1 February 2021)

Even fighting a lost cause, then, can be worthwhile. Many interviewees explain that it was awareness of collapse that had stimulated moral reflection and made them take important decisions, such as giving up an urban career, joining XR, or starting a small-scale ecological farm. One transitioner stresses that he is driven by the desire to struggle and work practically for others—'sacrificing time for the good of others'—because 'we have a hard time in front of us during the coming decades' (Swedish transition activist, interviewed 2 February 2021). A collapsologist similarly stresses the importance of 'being able to be the person I want to be in a crisis, and that is not a person who sits at home waiting for everything to blow over and eating canned food while his neighbours perish' (Swedish collapsologist, interviewed 9 August 2021). In this attitude, we sense the presence of a moral nature interest, in the sense that an awareness of nature's collapse stimulates a reflexivity that makes people wake up to their moral responsibility.

Many postapocalyptic environmentalists thus assert the value of struggling, even if it is for a lost cause—for dignity, for one's children, or simply for others. Often this is something that they *must* do to retain their sense of self-worth. Rupert Read, close to XR as well as to Deep Adaptation, states that we must act, for even if we fail, we will at least have tried: 'The thing that would be really intolerable, so shaming that it would be hard to square with life going on at all, would be if we reached (say) 2030... and we couldn't look our kids in the face' (Read & Alexander, 2019, p. 29). In a similar vein, Servigne and his co-authors write that the answer to fear is not hope or optimism, but courage (Servigne et al., 2021, p. 192). The fact that catastrophe is coming or is already progressing around them is, in their eyes, a cause not for cynical withdrawal but for moral engagement. Far from representing an easy way out, trying to be good here and now is precisely what is required in a world in which catastrophes have become unavoidable. Even if the struggle to avoid the collapse is doomed to failure, it is still morally meaningful, or even a duty, to do what one can to reduce suffering, injustices, and harm to nature. This call of morality is not necessarily seen as a burden but can also be presented as an uplifting and invigorating opportunity to prove one's worth. The Buddhist scholar Joanna Macy, for instance, writes about how lucky she feels to be alive at a time when people are starting to take radical action, such as the resistance at Standing Rock. Even if the future looks bleak, she writes, we can still think: 'How lucky we are to be alive now—that we can measure up in this way' (Macy, 2014).

Loyalty to the Dead

One way in which moral reasoning can become central to motivating activism is through loyalty to the 'dead' or to the 'dying'. This moral idea is of import in postapocalyptic environmentalism since it directly links up with the experience of loss and practices of mourning and shows how such experiences and practices can be politically mobilizing. In a memorable phrase, Benjamin suggested that 'the image of enslaved ancestors' might be more important in spurring people to revolt than 'the ideal of liberated grandchildren' (Benjamin, 1977b, p. 258). Remembrance of the dead or the dying can be politicizing since it can strengthen bonds of identification and solidarity with victims. The validity of this suggestion is borne out by examples from the history of postapocalyptic environmentalism. In the Ashio struggle, mentioned at the start of this chapter, participants in the struggle on several occasions expressed a loyalty to or even identification with the 'dead', meaning victims of past and present injustices, as could be seen when they signed petitions with the expression 'the dead' (Tanaka, 2004, p. 222, 2005, p. 196). When the government in 1907 razed Yanaka village and deported most of its inhabitants to make room for a dam intended to gather up polluted water from the Ashio mine, the movement leader Tanaka Shōzō and his followers decided to stay in the village, living in shacks with a handful of remaining villagers. Despite knowing that there was little or no hope of saving the village, he continued living there, struggling for its survival, until his death in 1913. When asked by a Buddhist priest if he thought there was any chance of success, he answered that he could not tell, 'but can one abandon someone who's dying [...] just because you don't think he'll recover?' (quoted in Strong, 1995, p. 163). Loyalty or duty to the village thus seems to have motivated him more than any actual hope of success. This attitude did not express defeatism, but rather a determination to struggle. The dying village was a moral touch-stone since it was where Japan could face the consequences of its modern-ization most clearly. His loyalty was linked to the fundamental moral worth that he saw as embodied in the village and the relationship to nature that it represented, but which was now threatened with extinction by modern civilization.

The example of Yanaka—a quintessential sacrifice zone—shows how the identification with the dead or the dying can be linked to the demand for justice. As we recall, a similar appeal to justice was voiced in the Tribunals for the Rights of Nature. Many other examples can be cited of how activists have been motivated to continue a struggle by feelings of loyalty and deep emotional bonds to a non-human nature felt to be dying, as for instance when tree-sitters stay on in a forest despite knowing that the trees are 'already dead' (Krøijer, 2019, p. 165ff). An even clearer example can be found in the protests against the mercury poisoning in Minamata Bay in Japan in the 1960s, where protesters—many of whom were fishermen who had lost parents or other family members through the poisoning—impersonated the dead by dressing in white, carrying black flags expressing their rancor and calling themselves the 'dead citizens of Minamata' (Watanabe, 2017). By contrast, future generations were hardly ever mentioned. It is important to realize that this orientation towards the dead was not just a matter of private grieving, but also served as a way of politicizing the environmental issue and bringing forward a justice claim. Without justice—including redress for the wronged and remorse from the responsible—the spirits of the dead would not be appeased.

Permanent Catastrophe and Utopia

We want to finish by further looking into the temporality we see in the postapocalyptic narrative. In this narrative, the catastrophes around us are part of the normal run of modern capitalist civilization, rather than aber-rations or accidents. Instead of seeing the catastrophes as remediable through 'progress', modernity is grasped as what Benjamin described it as: a continuous or permanent catastrophe. In his iconic representation of this notion, history is a pile of debris, growing skywards before the eyes of an angel who wants to go back and repair what is broken but who is blown backwards into the future by a storm blowing from paradise (Benjamin, 1977b, p. 255)—a vision that implies a radical rejection of the belief in linear progress. This view of history offers a stark contrast both to the future-oriented optimism of the green progress narrative and to the apocalyptic narrative according to which future threats can still be avoided. In a critique of future-oriented apocalypticism, Evan Calder Williams has offered a vision of history that resonates with Benjamin's notion of permanent catastrophe, suggesting that we are already now living through a 'combined and uneven apocalypse', an apocalypse that is in progress but unevenly distributed. Paraphrasing a statement by William Gibson, Eddie Yuen similarly states that 'the catastrophe is already here, it's just not evenly distributed' (Yuen, 2012, p. 130 n2).

In this chapter we have tried to throw light on the implications of this stance for political action. A contentious point—which has served as a lightning rod to those who accuse postapocalyptic environmentalists of defeatism—is that such environmentalism no longer views catastrophe as preventable. Either it is already in progress or else people experience themselves as powerless to prevent it. But as we have seen, this by itself does not necessarily mean passivity or that 'nothing can be done'. Instead, catastrophe becomes internalized as a necessary point of departure for political action, including attempts to salvage what can still be saved and demanding redress and settling wrongs. We have tried to show how the postapocalyptic narrative informs and helps sustain activism by pointing to the morally meaningful action that still remains possible even when civilization is felt to be crumbling.

Firstly, we can see how the idea of ongoing catastrophes underpins a highly polemical politics in justice struggles. The International Tribunals for the Rights of Nature break with the orderly temporality of progress, instead evoking the temporality of continuous catastrophe. Condensed in the trope of the violation and suffering of Mother Nature, this catastrophe is not a future threat, as in the apocalyptic narrative, but a present reality in the sacrifice zones. Utopia is not a goal at the endpoint of linear time, but rather the arrival of justice that puts an end to the violation. Regarding time, the tribunals enact the arrival of justice, which, as in a flash, illuminates the ongoing catastrophe in a utopian light. It is a utopia reminiscent of the present orientation of early modern Chiliasm's Millennium, representing both the end of the world as we know it and the beginning of a new time, defined by peace and justice. As highlighted by Karl Mannheim in his *Ideology and Utopia*, the chiliasm of Thomas Müntzer (see Chap. 3) involved an idea of social change as revolution that was different from both the premodern understanding of revolution as a restoration of a social order and that of Enlightenment modernity as a future-oriented collective act, a means to achieve a new social order:

> Chiliasm sees the revolution as a value in itself, not as an unavoidable means to a rationally set end, but as the only creative principle of the immediate present, as the longed-for realization of its aspirations in this world. (Mannheim, 1954, p. 196)

Further, reflecting the unevenness of the ongoing catastrophes, the collective 'we' is not 'humanity' in the abstract but a translocally articulated 'we' that includes a solidarity with nature. In line with this, local activities and local struggles tend to be centred in the tribunals. The chiliastic nature of the rights appealed to in the tribunals means that the appeals are not so much a call for *extending* the existing legal frameworks as a call for rejecting and transcending such frameworks. The very fact that a mock tribunal is needed is both an accusation against and a verdict on these frameworks. Nature is brought into being as a wronged subject in a process akin to what the philosopher Jacques Rancière calls subjectivation, the coming into being of a new political subject that disrupts the existing political field.

Secondly, the seemingly more apolitical groups devoted to mourning or preparing for collapse are also not passive. This is true also of groups focusing on cultural activism, such as Dark Mountain. To them, the bleak time of gradual decay and collapse is illuminated by a weak Utopian light that shines on history from the 'no-place' where participants in these groups remember and mourn their losses and reflect on their own responsibilities and commitments. This external gaze is meant to help them see what really matters to them with greater clarity. Often this leads them to practical work on the local level where they try to live closer to nature and in greater independence from the crumbling world of civilization. In many cases, here a new hopefulness akin to radical hope is generated.

Thirdly, elements of the postapocalyptic narrative are also present in groups calling for a climate emergency, such as XR and Fridays for Future. As in justice struggles, these groups aim to politicize the climate issue. Hope in a successful outcome, however, appears to play a comparatively insignificant role in motivating activism. Instead, activists express anger or, remarkably often, ethical motives related to dignity, duty, guilt or the desire to be able to look one's children in the face. Ethical reasons can thus to some extent be a substitute for hope in motivating activism, by making participation seem valuable in itself, regardless of the outcome. To the extent that a utopia is discernible in this version of the narrative, it consists in the vision of society as an ethical community, living up to its responsibility and taking appropriate action in response to the climate emergency.

We may ask if utopia has no other role in the postapocalyptic narrative than that of a make-believe arrival of justice or a 'no-place' from which to view ourselves. Is it not also possible to imagine utopia in the form of an *end* to the series of catastrophes and of the agony of Mother Earth? If it is true that the permanent catastrophe is a product of capitalist modernity, then surely a better future should be possible by ending that system. To be sure, many postapocalyptic environmentalists admit the possibility of a better world on the far side of collapse and even suggest that we think of the collapse as an opportunity for a fundamental reshaping of society—but these hopeful gestures are usually vague and inarticulate, as befits radical hope.

CONCLUSIONS AND CRITIQUE

Disasters of various kinds—ranging from poisoned waterways to global warming—have accompanied capitalism throughout its development. The fact that the environmental movement has formulated an apocalyptic narrative for much of the post-war period, which has mainly referred to the need to avert *future* disasters should not obscure the fact that many environmental struggles have revolved around already occurring or ongoing disasters. Today as in the past, people protest not only out of fear for the future, but also out of anger at existing and ongoing catastrophes and in the demand for justice.

The postapocalyptic narrative exists in several variants. The most important dividing line is at the same time class-based and geographical, as it runs between currents rooted in the centre and periphery of the capitalist world economy, where the periphery includes large parts of the Global South and the rural areas of the Global North. The clearest difference between the currents in the core and those in the periphery concerns movement strategies and action repertoires. The movement in the core has either focused on cultural activism, building resilient local communities or calling for emergency action. By comparison, the movement in the periphery has waged a far more visible struggle for justice and been far more vociferous in pointing out the inherent destructiveness of the capitalist economic world order based on the exploitation of people and land. Emotionally, the postapocalyptic narrative in both the core and the periphery offers a sharp contrast to the optimism and modernist self-confidence of the green progress narrative as well as to the combination of fear and hope typical of the apocalyptic narrative.

The postapocalyptic narrative has stirred up significant critique and debate within the environmental movement in the 2010s. Criticism has most often been directed at the 'Northern' variant. Dark Mountain, collapsology, and Deep Adaptation have all been criticized for being defeatist and politically ineffective (see e.g. Charbonnier, 2019; Kingsnorth & Monbiot, 2009; Bendell, 2020), while the transition movement's rejection of open protest has frequently led it to being described as partaking in a depoliticized or 'post-political' environmentalist discourse (Kenis, 2018). For example, when Dark Mountain was introduced in Sweden in 2010, climate activists and human ecologists Rikard Warlenius and Andreas Malm (2012) argued that the movement's message that it is 'too late' undermined the possibilities for mass action. Further, they argued that the view—which they associate with Paul Kingsnorth—that 'the climate movement should stop marching and instead go to the mountains and listen to what the wind has to say to the soul', fails to recognize that the possibilities for such a withdrawal to an idyllic life close to nature is being seriously threatened by climate change, thus making the postapocalyptic narrative just as detached from reality as 'the technofix' of the green progress narrative. Together with Swedish transition activist David Jonstad and Dougald Hine (2012) replied to the criticism by arguing that the postapocalyptic narrative more than any other approach is based on realism: 'to recognize how lost our society is, is liberating. We can then stop pretending and confront reality on a realistic ground'. These are basically the frontlines of the debate on the postapocalyptic narrative. Its critics claim that its outlook on the world encourages passivity, while its proponents argue that effective action is only possible if we stop denying the fact that the catastrophe is already occurring. Contra this argument, critics argue that such realism is anti-utopian, meaning that action without hope lacks the capability of being transformative and thus, despite good intentions, functions to reproduce the social order that has brought disaster. We will return to this debate in the next chapter, in which we also present our own views regarding the (ir)relevance of the three movement narratives that we have analysed in this book.

References

Adorno, T. W. (1978). *Minima Moralia*. Verso.

Albrecht, G. (2006). Solastalgia. *Alternatives Journal, 32*(4/5), 34–36.

Aminzade, R., & McAdam, D. (2001). Emotions and contentious politics. In R. Aminzade, J. A. Goldstone, D. McAdam, E. J. Perry, W. H. Sewell, S. Tarrow, & C. Tilly (Eds.), *Silence and Voice in the Study of Contentious Politics* (pp. 14–50). Cambridge University Press.

Anderson, B. (2017). Emergency Futures: Exception, Urgency, Interval, Hope. *The Sociological Review, 65*(3), 463–477.

Baker, C. (2014, June 17). Welcome to the Planetary Hospice. *Open Democracy*. https://www.opendemocracy.net/transformation/carolyn-baker/welcome-to-planetary-hospice.

Balibar, E. (2012). Justice and Equality: A Political Dilemma? Pascal, Plato, Marx. In É. Balibar, S. Mezzadra, & R. Samaddar (Eds.), *The Borders of Justice* (pp. 9–31). Temple University Press.

Bendell, J. (2020, August 31). To Criticise Deep Adaptation, Start Here. *Open Democracy*. https://www.opendemocracy.net/en/oureconomy/criticise-deep-adaptation-start-here/

Bendell, J., & Read, R. (Eds.). (2021). *Deep Adaptation: Navigating the Realities of Climate Chaos*. Polity Press.

Benjamin, W. (1977a). Zentralpark. In S. Unseld (Ed.), *Illuminationen, Ausgewählte Schriften 1* (pp. 230–250). Suhrkamp.

Benjamin, W. (1977b). Über den Begriff der Geschichte. In S. Unseld (Ed.), *Illuminationen, Ausgewählte Schriften 1* (pp. 251–261). Suhrkamp.

Berger, J. (1999). *After the End: Representations of Post-Apocalypse*. University of Minnesota Press.

Bounds, A. M. (2021). *Bracing for the Apocalypse: An Ethnographic Study of New York's 'Prepper' Subculture*. Routledge.

Cassegård, C. (2014). *Youth Movements, Trauma, and Alternative Space in Contemporary Japan*. Global Oriental.

Cassegård, C., & Thörn, H. (2018). Toward a Postapocalyptic Environmentalism? Responses to Loss and Visions of the Future in Climate Activism. *Environment and Planning E: Nature and Space, 1*(4), 561–578.

Cassegård, C., Soneryd, L., Thörn, H., & Wettergren, Å. (eds.) (2017). *Climate Action in a Globalizing World: Comparative Perspectives on Environmental Movements in the Global North*. Routledge.

Charbonnier, P. (2019). Splendeurs et misères de la collapsologie. Les impensés du survivalisme de gauche. *Revue du Crieur, 2*, 88–95.

Cunsolo, A., & Ellis, N. R. (2018). Ecological Grief as a Mental Health Response to Climate Change-related Loss. *Nature Climate Change, 8*, 275–281.

Dark Mountain Project. (n.d.). The Dark Mountain Project: Home. Retrieved January 11, 2012, from http://www.dark-mountain.net/

de Moor, J. (2021). Postapocalyptic Narratives in Climate Activism: Their Place and Impact in Five European Cities. *Environmental Politics.* https://doi.org/1 0.1080/09644016.2021.1959123

de Moor, J., De Vydt, M., Uba, K., & Wahlström, W. (2020). New Kids on the Block: Taking Stock of the Recent Cycle of Climate Activism. *Social Movement Studies.* https://doi.org/10.1080/14742837.2020.1836617

de Moor, J., & Wahlström, M. (2019). Narrating Political Opportunities: Explaining Strategic Adaptation in the Climate Movement. *Theory and Society, 48,* 419–451.

Derrida, J. (1990). Force of Law: The 'Mystical Foundation of Authority'. *Cardozo Law Review, 11*(5–6), 920–1045.

Derrida, J. (2006). *Specters of Marx: The State of the Debt, the Work of Mourning and the New International.* Routledge.

Du Cann, C., Hine, D., Hunt, N., & Kingsnorth, P. (Eds.). (2017). *Walking on Lava: Selected Works for Uncivilised Times.* Chelsea Green Publishing.

Foster, G. A. (2016). Consuming the Apocalypse, Marketing Bunker Materiality. *Quarterly Review of Film and Video, 33*(4), 285–302.

Franzen, J. (2019, September 8). What If We Stopped Pretending?. *New Yorker.* https://www.newyorker.com/culture/cultural-comment/what-if-we-stopped-pretending?verso=true

Gamson, W. A., & Meyer, D. S. (1996). Framing Political Opportunity. In D. McAdam, J. D. McCarthy, & M. N. Zald (Eds.), *Comparative Perspectives on Social Movements: Political Opportunities, Mobilizing Structures, and Cultural Framings* (pp. 275–290). Cambridge University Press.

Hage, G. (2003). *Against Paranoid Nationalism: Searching for Hope in a Shrinking Society.* The Merlin Press.

Hamilton, C. (2017). *Defiant Earth: The Fate of Humans in the Anthropocene.* Polity Press.

Hine, D. (2012). The Dark Shapes Ahead. Retrieved January 18, 2018, from http://dougald.nu/work/the-dark-shapes-ahead/

Hine, D. (2017). How Climate Change Arrives. Retrieved January 18, 2018, from http://dougald.nu/work/how-climate-change-arrives/

Hine, D. (2020). The Dark Matter of Climate Change. Retrieved November 26, 2020, from https://vimeo.com/474573744

Hine, D. (n.d.). There Is a Task Ahead of Us—To Regrow a Living Culture. *Dougald Hine.* Retrieved January 18, 2018, from http://dougald.nu/

Jonstad, D. & Hine, D. (2012) Vi hoppas inte på en klimatkollaps. *Aftonbladet,* June 1.

Kaplan, A. (2016). *Climate Trauma: Foreseeing the Future in Dystopian Film and Fiction.* Rutgers University Press.

Kenis, A. (2018). Post-politics Contested: Why Multiple Voices on Climate Change Do Not Equal Politicisation. *Environment and Planning C: Politics and Space, 37*(5), 831–848.

Kingsnorth, P. (2010, April 29). Why I Stopped Believing in Environmentalism and Started the Dark Mountain Project. *The Guardian.* http://www.theguardian.com/environment/2010/apr/29/environmentalism-dark-mountain-project

Kingsnorth, P. (2017). *Confessions of a Recovering Environmentalist and Other Essays.* Faber & Faber.

Kingsnorth, P., & Hine, D. (2017). *Uncivilisation. In P. Kingsnorth, Confessions of a Recovering Environmentalist and Other Essays* (pp. 257–284). Faber & Faber.

Kingsnorth, P. & Monbiot, G. (2009, August 17). Is There Any Point in Fighting to Stave Off Industrial Apocalypse?. *The Guardian.* https://www.theguardian.com/commentisfree/cif-green/2009/aug/17/environment-climate-change

Klein, N. (2014). *This Changes Everything: Capitalism vs. the Climate.* Simon & Schuster.

Kleres, J., & Wettergren, Å. (2017). Fear, Hope, Anger, and Guilt in Climate Activism. *Social Movement Studies, 16*(5), 507–519.

Krøijer, S. (2019). Slow Rupture: The *Art of Sneaking* in an Occupied Forest. In M. Holbraad, B. Kapferer, & J. F. Sauma (Eds.), *Ruptures: Anthropologies of Discontinuity in Times of Turmoil* (pp. 157–173). UCL Press.

Lear, J. (2006). *Radical Hope: Ethics in the Face of Cultural Devastation.* Harvard University Press.

Macy, J. (2014). It Looks Bleak. Big Deal, It Looks Bleak. Exopermaculture.com (posted April 2, 2014). Retrieved February 6, 2021, from https://www.exopermaculture.com/2014/04/02/joanna-macy-on-how-to-prepare-internally-for-whatever-comes-next/

Mannheim, K. (1954). *Ideology and Utopia: An Introduction to the Sociology of Knowledge.* Routledge & Kegan Paul.

Martinez-Alier, J. (2003). *The Environmentalism of the Poor: A Study of Ecological Conflicts and Valuation.* Edward Elgar Publishing.

Matsumoto, E. (2000). *Kōdokuchi no sanjo.* Yumani shoten.

Nixon, R. (2011). *Slow Violence and the Environmentalism of the Poor.* Harvard University Press.

Read, R., & Alexander, S. (2019). *This Civilization is Finished: Conversations on the End of Empire—And What Lies Beyond.* Simplicity Institute.

Richardson, M. (2018). Climate Trauma, or the Affects of the Catastrophe to Come. *Environmental Humanities, 10*(1), 1–19.

Rosa, H. (2019). *Resonance: A Sociology of Our Relationship to the World.* Polity Press.

Ruong, I. (1967). En utmaning. *Samefolket, 11–12,* 210–214.

Schlosberg, D., & Collins, L. B. (2014). From Environmental to Climate Justice: Climate Change and the Discourse of Environmental Justice. *WIREs Climate Change, 5*(May/June), 359–374.

Scranton, R. (2015). *Learning to Die in the Anthropocene: Reflections on the End of a Civilization.* City Lights Books.

Servigne, P., & Stevens, R. (2020). *How Everything Can Collapse: A Manual for Our Times.* Polity Press.

Servigne, P., Stevens, R., & Chapelle, G. (2021). *Another End of the World is Possible: Living the Collapse (and Not Merely Surviving It).* Polity Press.

Stephenson, W., & Kingsnorth, P. (2012, April 11). 'I Withdraw': A Talk with Climate Defeatist Paul Kingsnorth. *The Grist.* http://grist.org/climate-energy/i-withdraw-a-talk-with-climate-defeatist-paul-kingsnorth/

Stoner, A. M., & Melathopoulos, A. (2015). *Freedom in the Anthropocene: Twentieth-century Helplessness in the Face of Climate Change.* Palgrave Macmillan.

Streeby, S. (2018). *Imagining the Future of Climate Change: World-Making through Science Fiction and Activism.* University of California Press.

Strong, K. (1995). *Ox Against the Storm: A Biography of Tanaka Shozo—Japan's Conservationist Pioneer.* Japan Library.

Stuart, D. (2020). Radical Hope: Truth, Virtue, and Hope for What Is Left in Extinction Rebellion. *Journal of Agricultural and Environmental Ethics, 33,* 487–504.

Tanaka, Shōzō (2004). *Tanaka Shōzō bunshū.* Vol. 1 (*Kōdoku to seiji*). M. Yui, & H. Komatsu (Eds.). Iwanami shoten.

Tanaka, Shōzō (2005). *Tanaka Shōzō bunshū.* Vol. 2 (*Yanaka no shisō*). M. Yui, & H. Komatsu (Eds.). Iwanami shoten.

Tasset, C. (2019). Les 'effondrés anonymes'? S'associer autour d'un constat de dépassement des limites planétaires. *La Pensée écologique, 1*(3), 53–62.

Thunberg, G. (2019). *No One is Too Small to Make a Difference.* Penguin.

Utsi, P., & Utsi, I. (1980). *Giela Gielan.* Luleå alltryck.

Warlenius, R., & Malm, A. (2012). *'Dark Mountain' hoppas på en klimatkollaps. Aftonbladet.*

Watanabe, K. (2017). *Shimin to nichijō—Watashi no Minamata-byō tōsō. Genshobō.*

Westwell, E., & Bunting, J. (2020). The Regenerative Culture of Extinction Rebellion: Self-care, People Care, Planet Care. *Environmental Politics, 29*(3), 546–551.

Wilt, J. (2014). Admit Defeat, Then Engage. *Geez Magazine, 35,* 38–44. Retrieved January 25, 2018, from https://geezmagazine.org/magazine/article/admit-defeat-then-engage/.

Yuen, E. (2012). The Politics of Failure Have Failed: The Environmental Movement and Catastrophism. In S. Lilley, D. McNally, & E. Yuen (Eds.), *Catastrophism: The Apocalyptic Politics of Collapse and Rebirth* (pp. 15–43). PM Press.

CHAPTER 5

Towards a Critique of the Environmental Movement

Abstract Here we formulate a critical analysis of the environmental movement, returning to the question why it has failed to halt the environmental degradation. Inspired by the early Frankfurt School, we critically examine the limitations and possibilities of the three narratives. We argue that the environmental movement has failed not only because of capitalist resistance, but also because the movement has been shaped by capitalism in its thinking and modes of action. At the same time, all three narratives offer resources for understanding our predicament. The postapocalyptic narrative in particular is helpful because of its clear recognition that catastrophes are already ongoing and that the struggle must be about justice for the human and non-human victims just as much as about preventing further future disasters.

Keywords Environmental movement • Narrative • Critical theory • Catastrophe • Critique of ideology • Capitalism • Utopia

Throughout the book, we have repeatedly pointed to a paradox encapsulated in the notion of the Anthropocene. On the one hand, humanity has become more powerful than ever, but on the other, this very development brings us to the brink of destruction. Instead of liberating us, the

© The Author(s), under exclusive license to Springer Nature Switzerland AG 2022
C. Cassegård, H. Thörn, *Post-Apocalyptic Environmentalism*,
https://doi.org/10.1007/978-3-031-13203-2_5

development of productive forces has led to a new captivity, this time in a second nature compulsively geared to endless growth and capital accumulation, processes that since their inception have caused a cascading series of catastrophes for the most vulnerable human populations as well as many non-human life forms. This development could have been called blind if industrial capitalism had not from its very start been accompanied by critical voices and protest movements. The depletion of resources and derangement of the planetary ecology has not occurred due to ignorance. On the contrary, it has gone hand in hand with an increasing understanding of the disastrous consequences of expansion. If the spread of environmental consciousness today is the most tangible success of the environmental movement, its greatest failure is that the catastrophes have continued despite this environmental awareness. How can this failure be explained? What has made the system so seemingly immune to environmental criticism?

Using the concept of nature interest, we have sought to analyse the origins and development of the environmental movement in relation to the development of capitalism. Our answer to the above questions can be formulated by paraphrasing Marx (2010, p. 5): the environmental movement and its participants have contributed to shaping modern history, but not as they have pleased, not under conditions of their own choosing, but under circumstances transmitted from the past. Our suggestion is that the environmental movement—like so many other social movements fighting the ills of capitalism—has failed not only because capitalism has been an overpowering historical force, but also because capitalism has fundamentally shaped the movement, its social base, ideas, and strategies of action. Capitalism does not just involve the exploitation of natural resources; it shapes how we in modern society think about, and organize, the relationship between society and nature. Environmentalist ideas about nature—as a resource, a romanticized other, a complex ecosystem encompassing humanity, or a second nature created by us—reflect categorizations that have been functional to the capitalist system. Similarly, structural positions in the system are reflected in the movement's strategies and social composition, whether we look at currents rooted in bourgeois milieus or among the poor at the system's periphery.

All this helps to explain why the movement has had difficulty negating capitalism. But it is far from a complete explanation for the failure of the environmental movement. On its own, such an explanation risks subscribing to a historical determinism which denies people the ability to exercise their free, collective will. The failure of the environmental movement does not just reflect the straitjacket imposed by capitalist forms and categorizations on its thinking; it also speaks to the misguidedness and inadequacy of its collective will, manifested in its strategies and ideas.

This opens up the way for a critique of environmentalism. So far, we have devoted ourselves primarily to describing and analysing the history of the environmental movement, without taking a position on the ideas and strategies formulated and debated within it. In this concluding chapter, we want to step forward as critics of the environmental movement. We certainly share the premise of the environmental movement as it has developed over its long history, namely that the instrumental nature interest must be combatted and environmental destruction stopped. We add, however, that this struggle needs to be anti-capitalist. To stop the catastrophes, the destruction, the exploitation, and the degradation, a break with the capitalist system is required that clears a path for a more just and decent world.

'The tradition of all dead generations weighs like a nightmare on the brains of the living', Marx (2010, p. 5) famously asserted. The task of critique must be to dispel this nightmare. In this task we are inspired by Marx as well as by thinkers of the early Frankfurt School such as Walter Benjamin, Theodor W. Adorno, and Herbert Marcuse. Benjamin's critical analyses of the history of modernity aimed precisely at an 'awakening from the nineteenth century', or in other words, from the 'dream-filled sleep' that had settled over Europe with capitalism (Benjamin, 1999, p. 391). Like him, we hope for history to contribute to such an awakening—if not from the dreams of nineteenth-century Europe, then at least from those that have accompanied the history of the environmental movement. Although this movement is shaped by the development of capitalism, it is not entirely caught up in the latter. To the extent that movements are genuinely driven by some form of pain, shock, grievance or concern about justice, they are in touch with the contradictions generated by the system. Such feelings have the capacity to disrupt transmitted categorizations and offer a constant reminder of the inadequacy of the ideas and modes of action which the movement has inherited from the system. In that sense they offer an Ariadne's thread that can help it to extricate itself from the latter and contribute to a more effective criticism.

Fundamental to the critical procedure of the early Frankfurt School was that it combined relentless ideology critique with a concern for utopia, for the possibilities inherent in the criticized object. Below, we explain why we believe utopian thinking is essential to a reinvigorated environmentalism. We then turn to a critical analysis of the limitations and possibilities of the three narratives. We point out how they have all, in different ways, failed to negate essential features of capitalism and show how they all carry with them the seeds of their own distinctive varieties of depoliticization. At the same time, we stress that they all offer important resources for thinking about our present situation. In particular, we point to the contributions of the postapocalyptic narrative, because of its clear recognition of the insufficiency of existing institutions and established ideologies and because of its readiness to acknowledge that catastrophes are already here and that the necessary struggle is therefore not only about preventing further catastrophes, but also about justice and support for human and non-human victims.

UTOPIA AND CATASTROPHE

In Chap. 3 we cited Fredric Jameson's quip that today 'it is easier to imagine the end of the world than to imagine the end of capitalism' (2003, p. 76). While there is some truth in this drastic formulation, we may ask if it is not too pessimistic about the ongoing attempts to imagine and create a society beyond capitalism. The notorious pessimism of early critical theory should not be taken as resigned, but as radical and revolutionary—as a critical tool that through its very blackness seeks to call forth the utopian imagination that it ostensibly denies (Cassegård, 2021, p. 69f; Gunderson, 2021, p. 154). Recognizing good grounds for pessimism is not tantamount to resignation. We agree with Antonio Gramsci (1971, p. 175), who, even behind prison bars, believed that 'pessimism of the intellect' had to be combined with 'optimism of the will' through perseverance and active waiting. Despite the veracity of Jameson's statement, we therefore, like many climate justice activists across the world, believe that it *is* possible to imagine a radically different, post-capitalist future, in which a reconciliation between humankind and the rest of nature is possible. A utopian imagination is not a scientific prediction, yet nor is it a mere daydream; it is a subversive fantasy about what reality could be like, and as such it has an important role to play both when we criticize society and when we seek guidance for our action. Such an imagination can take many different forms. To Adorno, a 'true society' would be free from the compulsion to accumulate:

Perhaps the true society will grow tired of development and, out of freedom, leave possibilities unused, instead of storming under a confused compulsion to the conquest of strange stars. A mankind which no longer knows want will begin to have an inkling of the delusory, futile nature of all the arrangements hitherto made in order to escape want, which used wealth to reproduce want on a larger scale. (Adorno, 1978, p. 156f)

What is this 'true society', we may ask, if not the exact opposite of the human situation in the Anthropocene, when systemic constraints virtually ensure that not a single possibility for further growth can be left unused? To Adorno, the image of freedom was not development, but its opposite: '*Rien faire comme une bête*, lying on water and looking peacefully at the sky, "being, nothing else, without any further definition and fulfilment"' (ibid., p. 157).

Statements such as these must be seen in the light of the hope, as Adorno and his co-author Max Horkheimer state in *Dialectic of Enlightenment*, for an enlightenment which itself has become enlightened and therefore no longer obsessively strives to master nature. Only when humankind is freed from this compulsion can it recognize nature as a subject rather than as a mere resource to exploit. At the same time, it becomes free to recognize itself as nature rather than suppressing its natural side. Adorno's most famous example of how glimpses of such a mimetic approach to nature are possible even today is probably his description of a person who happens to meet the gaze of a wounded animal. 'The possibility of pogroms', he writes in *Minima Moralia*, 'is decided in the moment when the gaze of a fatally-wounded animal falls on a human being' (1978, p. 105). In such a situation, we may feel a mimetic impulse, an identification with the helpless animal in which we recognize it as a subject that should not suffer. But we can also suppress the impulse, harden our hearts, and reinforce the separation between ourselves and 'mere' nature. It is the latter option that has been institutionalized in today's society, where the modern self is shaped through increasing control over one's own inner nature, and where external nature—including unwanted human 'others'—is objectified into an expendable resource. It is when we choose this latter path that the 'possibility of pogroms' opens up, as Adorno drastically puts it (ibid., p. 105; see also Cassegård, 2021, p. 57; Cook, 2006).

Mimesis not only implies the possibility of a reawakened moral perspective on nature—as highlighted in Adorno's example of the tormented animal—but can also open our eyes to aesthetic qualities that are obscured in capitalist society. We have seen that the aesthetic nature interest has often been linked to a bourgeois sensibility, but it also carries a subversive potential that can be realized when taken as a starting point for a mimetic approach where nature is recognized as a subject. The critical theorist who most clearly highlights this potential is probably Herbert Marcuse, known for his catchphrase that 'nature, too, awaits the revolution' (1972, p. 74). The expression appears in his *Counter-Revolution and Revolt* from 1972, a work that reflects his appraisal of North American environmental activism at the time and in which he describes the coming into being of a new sensibility that he hopes will lead to new ways of interacting with nature, to a new science, and to a new technology. The struggle with nature, he suggests, may then 'subside and make room for peace, tranquillity, fulfillment. In this case, not appropriation but rather its negation would be the nonexploitative relation: surrender, "letting be", acceptance' (ibid., p. 69).

The philosopher Alfred Schmidt notes that Marcuse's thoughts align with those of the materialist thinker Ludwig Feuerbach, who had argued that it was the aesthetic outlook—in which we appreciate objects through our senses without concern for the benefit they have for us—that enabled nature to appear to us as subject. Through our senses of hearing, sight, smell, taste, and touch, nature regains the ability to speak to us that it loses when we intellectualize it by turning it into a mere idea. Feuerbach had argued that while thinking is intolerant, the senses allow nature to be what it is, namely a subject in its own right. Schmidt suggests that this is why the senses play a crucial role in Adorno and Marcuse's thoughts on a possible non-exploitative approach to nature. When we allow nature to speak to us, it ceases to be an object for us to unilaterally master. Instead, the interaction with nature stimulates our thinking, which thereby becomes formed in a dialogue with our sensuous experiences (Schmidt, 1973, pp. 31f, 45ff).

When utopias are dismissed as naïve or unrealistic, their essential function is overlooked. As we have pointed out, the function of utopia is primarily critical, to highlight contradictions and thereby facilitate an orientation in the present. Even if we cannot predict the future, we know that radical social transformation requires collective willingness to change and action. This presupposes utopian forms of action (see Chap. 1), that is, collective action that radically negates the present. It is when we see the present in the light of how radically different it could be that the negating power of utopia appears. By pulling us out of taken-for-granted beliefs, it helps us see ourselves and our world in a new light, as Ricoeur points out.

When it fulfils this function, it can contribute to an awakening from the nightmare or 'dream-filled sleep' that Marx and Benjamin wrote about, thus freeing us from our paralysis.

For utopia to have this critical function, however, we must not turn our gaze away from the catastrophes, the exploitation, and the oppression generated by capitalism. The overwhelming power and destructiveness of this system makes the dream of a world without catastrophes and injustices appear naïve and out of touch with reality, but this catastrophic reality also breeds the utopias and makes them necessary. Here it is worth recalling that the utopian thinking of the Frankfurt School was rooted not in any faith in linear progress, but in a rejection of it. This is evident when Benjamin and Adorno describe modernity as a 'continuous' or 'permanent catastrophe'. For them, modernity was an era that incessantly shattered people's dreams and aspirations. Solidarity with the victims was only possible by recognizing their suffering, which should not be legitimized on the grounds of progress or historical necessity. For Benjamin (1996, p. 402), it was not the train of world history that offered salvation from the catastrophes, but rather the 'emergency brake'. Similarly, for Adorno it was the simple demand that suffering must end that best expressed solidarity with the victims, not the idea of progress as conventionally understood. Only by interrupting the catastrophic present could real progress become possible, for the first time in history (Adorno 1973, pp. 320, 2005, pp. 143–160; Benjamin 1977, p. 255; 1999, p. 473). In this approach to catastrophes and suffering, there are remarkable similarities with some of the variants of the postapocalyptic narrative that we have seen. For example, the idea of putting an end to suffering and interrupting the catastrophic course of real history brings to mind how the arrival of justice is pictured in the International Tribunal for the Rights of Nature (see Chap. 4).

In Benjamin and Adorno, the image of modernity as an ongoing series of catastrophes coexisted with the hope that these catastrophes would bring about an awakening. Out of pain, dizziness, and shock, impulses to criticize the dream world of capitalism could be born, albeit at first only in the form of a vague notion that 'something is missing' or that 'this cannot be everything' (Adorno, 1975, p. 368; Adorno & Bloch, 1988, p. 1ff). Similarly, we believe that a clear focus on present catastrophes is needed to think beyond capitalism. Rather than downplaying them in favour of uplifting narratives, it is up to the environmental movement to embrace the dissonance that catastrophes create and utilize them to highlight that capitalism systematically creates problems that it cannot solve. At the same time, we absolutely reject the idea that catastrophes on their own will bring about the end of capitalism. Capitalism has historically shown an outstanding ability to survive crises, and as long as at least some capitalist actors can continue to accumulate capital, nothing prevents a capitalist logic from operating in a world dragged down by catastrophes. Plenty of examples show how crises and widespread destruction have created new profit opportunities for capitalism. As Naomi Klein (2008) observes, capitalist elites have often skilfully exploited catastrophes to push through agendas that would otherwise likely have encountered fierce popular resistance. However, precisely to avoid more triumphs of such 'disaster capitalism', we must not be disoriented and unprepared when disaster strikes, and this in turn requires that we give catastrophes the attention they deserve already now. Only then will it be possible to organize the resistance and create the bonds of solidarity that will be needed to prevent disasters from being exploited by capitalism. In other words, what is needed is not a passive reliance on catastrophes to undermine capitalism, but an *interplay* between catastrophes and anti-capitalist movements, in which the movements actively intervene to turn catastrophes into a critique of the system.

Limitations of the Three Narratives

The fact that we emphasize the reality of catastrophes means that we share some ground with the postapocalyptic narrative. This does not mean that we do not also see grains of truth in the green progress narrative and the apocalyptic narrative. To the green progress narrative, however, we are more sceptical than to the other two. To the extent that it highlights the *automaticity of* progress, this narrative risks legitimizing the established system and passivating people. When nature is to be saved by being turned into capital, or when environmental problems are expected to be solved by economic growth, environmentally conscious consumers, and an economization of nature involving market mechanisms such as emissions trading, then the need for an environmental *movement* disappears since the system becomes viewed as capable of reforming itself through business-as-usual. At most, the role of the environmental movement would be to assist the system's own self-correction mechanisms by disseminating information to the public and providing expertise to companies and authorities.

The passivizing effect that occurs when progress is taken for granted as a feature of the system is *one* kind of depoliticization that can be discerned in the environmental movement. It has been especially conspicuous among environmental organizations that are institutionalized or that aspire to institutionalization. Such organizations may choose to downplay conflict for the benefit of a technocratic approach to environmental conflicts, or in order to preserve good relations with political institutions and 'green business'. This form of depoliticization can be pragmatically justified to the extent that the gradualist approach associated with the green progress narrative is capable of effectively dealing with environmental problems, but it can also fetter the movement if it prevents the latter from recognizing the truth of other narratives.

To a large extent, the institutionalization of the environmental movement that took place in the Global North during the latter part of the twentieth century has resulted in many environmental organizations settling into this limited role, something that can be seen in their embrace of the economic nature interest underlying emissions trading. It may be true that institutionalization has led to these organizations gaining easier access to power, but the major problem with this movement strategy is that it is based on an unrealistic premise: namely that a sustainable society is possible within the framework of business-as-usual. Here the seemingly 'realistic' adherence to the status quo reveals itself as wishful thinking. Those who advocate business-as-usual have so far utterly failed to show how a system that depends for its survival on constant growth and accumulation of capital can reduce greenhouse gas emissions at the rate needed to meet the pledges of the Paris Agreement. The assumption that a miraculous technology, non-existent today, will suddenly materialize and save the day is hardly convincing. Equally serious is that many of the measures taken against climate change, such as a drastic increase in electrification, themselves pose significant environmental risks in terms of resource extraction and the destruction of biotopes, not to mention social ills such as new sacrifice zones and deplorable labour conditions for miners and other workers. A miraculous technological fix against climate change could very well aggravate these environmental and social harms by giving the system, with its compulsive drive to incessant growth, a new lease on life.

We believe that the environmental organizations that embrace business-as-usual have fatefully prioritized what appears politically possible within the framework of the current system over what is required to achieve the climate and other environmental goals that they claim to represent. We agree with Greta Thunberg's remark—quoted in Chap. 4—that such political pragmatism is inadequate when it comes to nature. Today's crises, as has been pointed out, require us to redefine what realism means. Since conventional realism has become inadequate and disconnected from reality, only 'unrealistic' demands that go beyond the limits of what the system can accommodate offer a realistic way of dealing with these crises (Rosewarne et al., 2014, p. 5).

However, as we have seen, there are also socialist variants of the green progress narrative that challenge the established system. Here, the promise of technology is usually combined with an understanding that popular mass mobilizations are needed to break the resistance of the social elites and for the necessary transformation to happen with sufficient speed. The clearest example is probably the more radical variants of the Green New Deal that have emerged in the United States. We find much to affirm in this political programme, especially the strong presence of a social nature interest with an emphasis on a justice perspective, which is indispensable in an anti-capitalist environmental movement. At the same time, our reservations about the progress narrative apply to this programme too. When criticism of capitalism is downplayed—as it is when representatives of the programme explicitly claim that it is first and foremost a matter of saving the climate and that this criticism must therefore come second—and when the programme seemingly allows for all available technological means to dismantle the fossil fuel economy as quickly as possible—including not only renewable energy but also nuclear power and geoengineering (see Chap. 2)—then the difference with conventional ecomodernism risks being erased.

Perhaps the programme's reliance on the promise of increased prosperity and abundance is its most serious weakness (even when this promise is qualified into the formula of 'public abundance' and 'private sufficiency'). One does not have to be a prophet of doom to admit that today's waste of resources and environmental degradation may well land us in a future in which humanity is forced to live under greater scarcity and hardship than today. Insisting that a green socialist mobilization requires a promise of increasing prosperity can be dangerously counterproductive. If we really think that such a promise is required, what do we do if the collapse comes? In a recent work, *Affluence and Freedom*, the French philosopher Pierre Charbonnier (2021) argues that modern political ideas of freedom and autonomy have always rested implicitly or explicitly on the premise of affluence, and he raises the question of what political idea of freedom is possible when abundance is no longer possible. Is not this the question that we ought to consider, rather than taking for granted that affluence is a *sine qua non* for freedom? Should we not, in other words, aim for a strategy that makes it possible to work for a socialist utopia of justice, fair distribution, democracy, and mutual aid that would be viable *even* in view of collapse or an end of affluence? Furthermore, it can be questioned on empirical grounds whether the best way to mobilize a mass movement really is to promise material prosperity. As the protests in Ashio and Minamata show, mass movements can emerge without this promise. Rather than affluence or progress, it was justice and the desire to settle past and present wrongs that were central to these protests. What was effective in mobilizing support was that participants felt that they had right on their side and that their action could put an end to an injustice.

The apocalyptic narrative is, in our eyes, more justifiable than the narrative of green progress—both in that it clearly recognizes the possibility of a catastrophic development and through its view of history as contingent. Even as it gesticulates towards the apocalypse, it leaves room for people to actively choose a better future. Rather than passivating people, as the narrative of progress tends to do, it is usually animated by a critical impulse and a desire to break with the development towards catastrophe. The vision of an apocalyptic future highlights the need to protect nature's aesthetic values and ecological integrity. The aesthetic and ecological nature interests can thus be given a system-critical expression when incorporated into this narrative. As the history of the environmental movement in the post-war period shows, this narrative allows for an emphasis on a social nature interest and has at times even inspired a powerful critique of capitalism. This has happened when there has been a recognition that the apocalypse is caused by capitalism and that it is only by abolishing the latter that the apocalypse can be avoided.

Yet, the fact that an apocalyptic narrative in the twenty-first century has also in some cases been embraced by establishment actors shows that this narrative is also compatible with a defence of capitalism, given that markets and technology are seen as sufficient to avert the looming apocalypse. The ease with which sections of the establishment have been able to absorb this narrative may, paradoxically, be a consequence of the fact that the narrative tends to focus precisely on the threat of doom. The goal of avoiding the apocalypse can easily be deflected into a defence of the status quo. Despite the narrative's critical starting point, it then takes on a system-preserving function. Here we see another form of depoliticization—next to the passivizing effect noted above. According to the critics of the apocalyptic narrative that we discussed in Chap. 3, the narrative has become part and parcel of a post-political framing of the climate. Not only does it deflect criticism from the present system, but it also obliterates differences between perpetrators and victims through its reference to a universalized 'we' or 'everybody', standing for humanity in the abstract. This blindness to differences relating to class and social geography reinforces the narrative's tendency to neglect catastrophes that are already occurring and which it is therefore too late to avert. When catastrophes are placed solely in the future, the present is idealized into an idyll which the movement must protect against threats that are seen as extrinsic to the system itself. While the urge to protect and preserve is certainly often justified, such a movement risks overlooking that the present is already catastrophic for many people as well as non-human life forms, and that many catastrophes today are systematically produced by the system that defines the present, namely capitalism.

Connected to the above view is a further problematic aspect of the apocalyptic narrative, namely the ease with which it can slip into a justification of authoritarian technocratic interventions from above, paving the way for the development of a 'Climate Leviathan' that would allow capitalist elites to stabilize their positions and defend existing hierarchies amidst planetary environmental crisis (Wainwright & Mann, 2018). The risk of such a development is all the greater the more the apocalyptic narrative relies on a 'scientific' framing of the environmental crisis to the exclusion of broader social or justice-related concerns. Invoking the apocalypse is thus highly ambiguous. While on the one hand, it keeps history open, vouchsafes hope, and asserts the meaningfulness of action, on the other hand, it may function ideologically by legitimizing the status quo, promising to 'save' the latter while disregarding the injustices that inhere in it.

This brings us to the postapocalyptic narrative. As mentioned, we share this narrative's focus on the catastrophes that are already occurring and believe that it is only based on these that a system-critical utopian vision can be formulated. We also sympathize with the narrative because we see it as understandable and necessary that people react to these catastrophes through grief, mourning, and attempts at practical and mental reorientation and preparation. At the same time, we may question why it is that some of these activists only experience this grief now, as if capitalism hadn't always been shockingly catastrophic. The belatedness by which these activists have awakened to the arrival of catastrophe is reflected in the apolitical form that this narrative has often taken in the Global North which has made it easy to criticize it for defeatism. A further form of depoliticization makes its appearance in these groups, which results from the fact that they only partially operate in the public sphere, preferring to focus on practical activities such as permaculture or the mental reorientation needed to adapt to climate change. Of course, there have sometimes been good reasons for this apolitical stance. The political disillusionment that is discernible in several of these groups is understandable considering the inability of the political system to deal with today's most serious environmental crises. In many cases, such as the transition movement, the choice to operate in a seemingly apolitical fashion is a deliberate strategy, based on the conviction that prefigurative practical activities and 'constructive' forms of resistance (Sørensen, 2016) are a more efficient way to bring about change than by public protest. However, the tendency to downplay conflict risks making the postapocalyptic narrative harmless to the system causing the catastrophes. It thereby risks becoming complicit in 'sustaining the unsustainable' through merely 'simulative' behaviour that does little to stop injustices and environmental destruction (Blühdorn, 2017). Although constructive resistance is necessary, there are times when it is insufficient and instead confrontation is necessary. To clarify our point, we quote a young Japanese environmental activist, who, when we interviewed him in 2014, had long been a practitioner of permaculture, but who now saw a need for a confrontational form of activism in the wake of the Fukushima nuclear disaster:

> Even if all the people who are conscious moved out to the rural areas, that wouldn't solve the problems at all. Like, our issues are not rural issues. They're urban issues. Whether it's an agricultural problem or the nuclear problem or climate change, that money and those decisions are made in New York or Washington DC or Tokyo or Paris or, you know, London. These cities are the heart of darkness ... I can create my beautiful permaculture paradise in a rural area and then something's done in Tokyo or London or Washington DC and it goes off, becomes a toxic waste site or a power plant, and the dream's gone. If we don't change the urban environment, there's nowhere you can run. (Quoted in Cassegård, 2017, p. 161)

In other words, a strategy of mere withdrawal in order to build alternatives can be woefully insufficient, as the organic farmers in the Fukushima area experienced when the fallout of the disaster and public concerns about contamination came close to destroying their livelihoods.

It is important, we believe, not only to pay attention to the many catastrophes today, but also to diagnose their causes correctly. Some versions of the postapocalyptic narrative tend to naturalize collapses and catastrophes, seeing them as an inescapable feature of human history. It is certainly true that civilizations have always risen and fallen. It is also true that such a perspective on history can offer emotional relief and a sense of philosophical distance. Yet in some cases, it can also lead to a cynical view of suffering as inevitable and ridicule towards those who attempt to alleviate it. In conservative or far-right variants of the postapocalyptic narrative, it can form the basis of a 'lifeboat ethic' that would turn its back on climate refugees, opening the way for what Naomi Klein refers to as 'climate barbarism'. Climate change, she points out, 'isn't just about things getting hotter and wetter; under our current economic and political order, it's about things getting meaner and uglier' (Klein, 2020, p. 166). The acceptance of suffering as inevitable here becomes an excuse for inflicting it on others. This naturalization of catastrophes overlooks the fact that the emergence of capitalism is not a natural fact, but a historical process that has resulted in a qualitative difference between present and past historical catastrophes. Never have humans affected the conditions for life on Earth as directly and drastically as now, through climate disruption, resource exhaustion, pollution and species extinction. Today's catastrophes differ from premodern ones by being rooted in a global system that has been uniquely successful in expanding itself by plundering the earth of its resources and which now risks dragging the earth with it in its fall. Instead of naturalizing this system, we must bear in mind that it is a historical abnormality which will certainly not last for ever.

The postapocalyptic narrative does not have to be apolitical or defeatist. On the contrary, as the environmentalism of the poor shows, it can go hand in hand with struggle, radicalism, and powerful challenges to the system that causes the catastrophes. Acknowledging that catastrophes are already here need not in any way make the fight less radical. Solidarity with the victims or an identification with the dead can motivate activism as forcefully as the hope of averting future collapse. Attending to the environmentalism of the poor is also important because it offers what is probably the clearest system-critical formulations of the social and moral nature interests. It is in this current that we find the most emphatic demands for social justice, and the greatest readiness to accord a moral value to nature. In the Ashio struggle, it was precisely the reference to nature ('heaven') that allowed Tanaka Shōzō to denounce the immorality of industrializing Japan (Stolz, 2014). An equally clear example is the idea of the rights of nature as put forward among indigenous peoples—an idea that has inspired not only the Tribunal for the Rights of Nature, but also attempts to institutionalize these rights in the Global North that have led to what we have called the juridical interest in nature.

To counter the image of the postapocalyptic narrative as necessarily apolitical, we want to emphasize that almost *all* the currents that today challenge the institutionalized environmental movement are imbued to varying degrees by the realization that catastrophes are already a reality. This applies not only to culturally oriented activism of the type that emphasizes internal transition or 'deep adaptation', but also to large parts of the climate justice movement and the wave of climate protests in recent years. Among the latter we find XR and Fridays for Future, both of which stress that catastrophes are already here and that it is imperative to act immediately to avoid catastrophic global warming. Both groups express a desperate hope that such avoidance will still be possible, and in that sense their narrative is apocalyptic. At the same time, their narrative has postapocalyptic features due to the awareness of the devastation that has already started, and the associated emotional states of grief and despair. The interweaving of elements from the apocalyptic and postapocalyptic narratives reflects the fact that catastrophes are not only here, but can also get worse. In such a situation, it is perfectly understandable that activists both despair of what is happening and desperately try to avoid the worst.

Paradoxically, however, the very success of these groups in mobilizing a worldwide protest wave in 2019 may have been bought at the cost of still another form of depoliticization—one that easily occurs in broad, popular mobilizations and that is often particularly visible in single-issue movements when they attempt to gather supporters from all sides of the political spectrum. Here the best example might be XR, which explicitly aims to be a movement beyond right and left and for that reason concentrates all efforts to mobilize along a single conflict line at the expense of downplaying other conflict lines. While this is sometimes necessary in order to achieve a forceful mobilization, it produces a form of depoliticization as a side-effect. In the case of XR, this has left the group vulnerable to the charge of being insensitive to issues of class and race (Kinniburgh, 2020; Slaven & Heydon, 2020).

Clearly, the different forms of depoliticization that we have surveyed so far occur for different reasons—the desire to facilitate institutionalization, the wish to reach out to a universalized 'we', the attempt to mobilize a broad movement beyond left and right, and the choice to prioritize mental reorientation and constructive activism above confrontation. Although they can all appear justified depending on the situation, they all carry the risk of incapacitating the movement when it becomes necessary for it to engage in conflict.

BEYOND CAPITALISM

Criticism can thus clearly be directed at all three of the environmental movement's narratives. Each carries limitations that put it at risk of losing its system-critical potential and being co-opted to legitimize the industrial capitalism, with its compulsive accumulation of capital, that we believe is the single most important cause behind today's environmental problems. Yet criticism does not imply rejection; rather it aims as examining the limits of the validity of different perspectives and perceptions. Through our criticism we want to highlight the contributions that each narrative can make to overcoming capitalism.

For example, we welcome the measures to rapidly dismantle the fossil fuel industry, the massive redistribution of resources and the strengthened role of the labour movement advocated in socialist variants of the green progress narrative, but we are critical of the tendency to portray these measures as part of a linear progress wedded to economic growth. We are in favour of the apocalyptic narrative's emphasis on the crucial role of agency in the present, but critical of its narrow focus on merely future catastrophes. Finally, we appreciate the sharp eye of the postapocalyptic narrative for the suffering and catastrophes that are already unfolding and the attempts to build alternatives here and now, although we are critical of the apolitical tendency in some variants of this narrative.

With this criticism, we want to point towards what we believe would be a fertile point of departure for an anti-capitalist environmental movement that in recent years has been emerging under the umbrella of climate justice. We support a movement which, firstly, clearly recognizes the role played by industrial capitalism in creating today's environmental problems and aims for a revolutionary transformation of the economy. Secondly, this movement also recognizes that there is nothing automatic about this transformation, which presupposes political struggle. Thirdly, it acknowledges that the catastrophes are already here, and that the fight is therefore not only about preventing future catastrophes, but also about justice for the human and non-human victims who have already been affected.

Finally, we want to return to the problem that Jameson (2003, p. 76) raised: how do we imagine a world beyond capitalism? Before answering, we want to stress once again that utopias are not predictions or blueprints. The truth of a utopia does not lie in how well it predicts the future, but in how well it helps us orient ourselves in the present. How does utopia help us see the present and ourselves in a clearer light? How does it help us see what is at stake and discern opportunities for action? We have argued that utopias, to facilitate this orientation, must be based on the catastrophes that characterize our times. This is not just because utopias must answer to the desire for justice of those affected. Through the catastrophes, nature makes itself felt, by revealing the fragility of the categories and forms of thought through which we have sought to master it. In this way, catastrophes can prevent movements from becoming too swallowed up by the system. They can liberate our imagination and force us to think beyond capitalism.

At the same time, catastrophes present a particular challenge to utopian thinking. The destructiveness that has characterized capitalist development may well mean that a post-capitalist society must be built up in a world depleted of resources and in many ways more inhospitable than before to human life as well as to many other life forms. The utopia we imagine is therefore not characterized by material abundance. Yet we believe that it is a hopeful vision, for two reasons.

First, there is something hopeful, we think, in remembering that justice, grassroots democracy, mutual aid, and care of nature do not require material plenty. Throughout history, people have been willing to share burdens and help each other, as long as they have perceived the burden-sharing as fair and been treated with respect. Fairness and respect, we believe, may be more important than material abundance for wellbeing. Even though we are certainly not hopeful about maintaining the wasteful consumerist lifestyles of rich countries today, we are hopeful about humanity's ability to achieve and maintain a good post-capitalist society even without a high level of material prosperity.

Second, our utopia is a society in which neither humans nor non-human nature are sacrificed for the economic interests of elites. It is a society which is no longer subject to the compulsion to accumulate capital, and one which is therefore governed not blindly by economic processes but by people themselves in jointly made, democratic decisions. It is also a society where the relationship with nature is drastically different than today. When the compulsion to accumulate is suspended, the logic behind the accelerating destruction of nature since the beginning of industrial capitalism will also cease to operate. Although people must still work with nature, and in that sense relate instrumentally to it, it will not be regarded solely as a resource for sustaining human societies. Here is an opportunity for the liberation of nature and for the recognition of nature as a subject that Marcuse envisioned, and for the reconciliation with nature that Adorno hoped would occur when humanity recognized itself as nature.

Our vision, then, is of a socialism—in the sense of a free, democratic, and egalitarian society that does not sacrifice the weakest—that would be viable even without affluence. We do not advocate this vision out of love of austerity, but preparing for such a future is imperative, considering the likelihood of climate change and resource depletion leading to increasing hardships in the future. Not preparing for such a future would be irresponsible and counterproductive. We would be shooting ourselves in the foot if we were to claim that socialism would only be possible in affluent societies. That would make us helpless against climate barbarism. As catastrophes multiply and intensify, such barbarism may well gain ground. To us, climate barbarism is not just about outright xenophobia and eco-fascism, but stands for all action undertaken to *preserve privilege* in the face of climate change. For instance, when Clive Hamilton (2010, p. 218) warns that a likely development in view of global warming is that 'the ruthless and the wealthy use their power to control dwindling resources and exclude others from sharing in them', then that too is a form of climate barbarism. It is a reactionary response to climate change that instead of denying it concedes its reality and precisely for that reason opts for self-preservation while letting large swaths of humanity as well as non-human nature perish (Klein, 2020; Blumenfeld, 2022). To counteract that, we need to consider already now how to maintain a non-barbaric society in the event of a collapse or worsening catastrophes, so that we can act with decency and solidarity instead of turning on each other. We fully agree with Hamilton that mass mobilizations against inequality and exclusion are already needed to 'democratize survivability' (Hamilton, 2010, p. 223), or in other words to create societies that are not elitist, classist or bigoted and that will not sacrifice the poorest and most vulnerable when things get rough.

But how is the socialist utopia we propose to be realized? This is not the task for this book, but something that has to be shaped in the praxis of an anti-capitalist environmental movement. Let us just point out two things. The first is that even if things look bleak at present, there is an element of unpredictability in history that is a source of hope. That is why Benjamin, who was no stranger to dark broodings, urged us to act in the spirit of the fairy tale: 'with cunning and in high spirits' (2007, p. 102). The second point is connected to the first. Although our society may appear like an unchangeable second nature, it is in fact a historical creation that will change. Even in periods when the system appears stable, it is important to continue to criticize it and, to the extent possible, cultivate practical alternatives built on non-oppressive relations to humans as well as non-human nature that can be scaled up when the opportunity arises. To cunning and high spirits, then, we should add perseverance, an attitude that means not resignation, but rather an active and vigilant readiness. Precisely because we do not know the future, we need to be ready to grasp the chances that may come our way, no matter how small they may seem.

References

Adorno, T. W. (1973). *Negative Dialectics*. Routledge & Kegal Paul.

Adorno, T. W. (1975). *Negative Dialektik*. Suhrkamp.

Adorno, T. W. (1978). *Minima Moralia*. Verso.

Adorno, T. W. (2005). *Critical Models: Interventions and Catchwords*. Columbia University Press.

Adorno, T. W., & Bloch, E. (1988). Something's Missing: A Discussion Between Ernst Bloch and Theodor W. Adorno on the Contradictions of Utopian Longing. In *The Utopian Function of Art and Literature* (pp. 1–17). MIT Press.

Benjamin, W. (1977). Über den Begriff der Geschichte. In *Illuminationen, Ausgewählte Schriften 1* (pp. 251–261). Suhrkamp.

Benjamin, W. (1996). *Walter Benjamin: Selected Writings, 1938-1940* (M. Bullock & M. W. Jennings, Eds.). Harvard University Press.

Benjamin, W. (1999). *The Arcades Project*. The Belknap Press of Harvard University Press.

Benjamin, W. (2007). The Storyteller: Reflections on the Works of Nikolai Leskov. In H. Arendt (Ed.), *Illuminations* (pp. 83–109). Schocken books.

Blühdorn, I. (2017). Post-Capitalism, Post-Growth, Postconsumerism? Eco-Political Hopes Beyond Sustainability. *Global Discourse, 7*(1), 42–61.

Blumenfeld, J. (2022). Climate barbarism: Adapting to a wrong world. *Constellations, 1–17*. https://doi.org/10.1111/1467-8675.12596

Cassegård, C. (2017). Between Government and Grassroots: Challenges to Institutionalization in the Japanese Environmental Movement. In C. Cassegård, L. Soneryd, H. Thörn, & Å. Wettergren (Eds.), *Climate Action in a Globalizing World: Comparative Perspectives on Environmental Movements in the Global North* (pp. 149–169). Routledge.

Cassegård, C. (2021). *Toward a Critical Theory of Nature: Capital, Ecology, and Dialectics*. Bloomsbury.

Charbonnier, P. (2021). *Affluence and Freedom: An Environmental History of Political Ideas.* Polity.

Cook, D. (2006). Nature Becoming Conscious of Itself: Adorno on Self-Reflection. *Philosophy Today, 50*(3), 296–306.

Gramsci, A. (1971). *Selections from Prison Notebooks.* Lawrence & Wishart.

Gunderson, R. (2021). Hothouse Utopia: Dialectics Facing Unsavable Futures. *Zero books.*

Hamilton, C. (2010). *Requiem for a Species: Why We Resist the Truth About Climate Change.* Earthscan.

Jameson, F. (2003). Future City. *New Left Review, 21,* 65–79.

Kinniburgh, C. (2020). Can Extinction Rebellion Survive? *Dissent, 67*(1), 125–133.

Klein, N. (2008). *The Shock Doctrine: The Rise of Disaster Capitalism.* Penguin.

Klein, N. (2020). *On Fire: The Burning Case for a Green New Deal.* Penguin.

Marcuse, H. (1972). *Counter-revolution and Revolt.* Beacon Press.

Marx, K. (2010). The Eighteenth Brumaire of Louis Bonaparte. In *Marx Engels Collected Works* (Vol. 11, pp. 99–197). Lawrence & Wishart.

Rosewarne, S., Goodman, J., & Pearse, R. (2014). *Climate Action Upsurge: The Ethnography of Climate Movement Politics.* Routledge.

Schmidt, A. (1973). *Emanzipatorische Sinnlichkeit: Ludwig Feuerbachs anthropologischer Materialismus.* Carl Hanser Verlag.

Slaven, M., & Heydon, J. (2020). Crisis, deliberation, and Extinction Rebellion. *Critical Studies on Security, 8*(1), 59–62.

Sørensen, M. J. (2016). Constructive Resistance: Conceptualising and Mapping the Terrain. *Journal of Resistance Studies, 2*(1), 49–79.

Stolz, R. (2014). *Bad Water: Nature, Pollution, and Politics in Japan, 1870-1950.* Duke University Press.

Wainwright, J., & Mann, G. (2018). *Climate Leviathan: A Political Theory of Our Planetary Future.* Verso.

Index[1]

A

Accelerationism, 46
Adorno, Theodor W., 15, 48, 93,
  117–119, 130
Agenda 21, 43
Animal rights movement, 31, 40, 87
Anthropocene, 2, 16–18, 48, 64, 72,
  84, 113, 117
Ashio Copper Mine, 78, 83

B

Bakhtin, Mikhail, 10
Bastani, Aaron, 45, 46
Battistoni, Alyssa, 46, 47
Benjamin, Walter, 13, 14, 16, 17, 79,
  89, 103–105, 115, 119, 130
Biodiversity, 84
Borgström, Georg, 57–64, 60n1, 66,
  68, 69, 79
Brundtland, Gro Harlem, 43

C

Capitalism, 1, 4, 5, 9, 13–15, 17, 20,
  22, 30, 33, 38, 39, 41, 44, 47,
  48, 56, 58, 59, 72, 79, 90, 107,
  114–116, 119, 120, 122–124,
  126, 128–131
Capitalocene, 17, 72
  See also Anthropocene
Carbon emission, 49
Carbon trading, 44, 120, 121
Carson, Rachel, 2, 54, 56–58, 61–67,
  72, 78, 79
Catastrophe, v, 2–5, 24, 45, 57, 60–62,
  74, 78–80, 84, 85, 97–99, 102,
  104–108, 114, 116–120, 123
  continuous/permanent, 104
Certainty, 29–31
Charbonnier, Pierre, 108, 122
Chiliasm, *see* Millennialism
Class, 19, 21–23, 34, 41, 46, 66, 73,
  124, 127

---

[1] Note: Page numbers followed by 'n' refer to notes.

Climate barbarism, 126, 130
Climate justice, 9, 69, 71–74, 81, 84, 88, 95, 116, 127, 128
Climate movement, *see* Environmental movement
Club of Rome, 49, 58
Collapsology, 94, 95, 108
Collective identity, 4, 6, 7, 11, 19
Commoner, Barry, 16, 54, 57, 61, 64–70
Communist Manifesto, 54, 61, 62
Conservation movement, 5, 19, 21, 23, 30–32, 32n1, 36–38, 40–42, 49, 55, 93
Copenhagen COP meeting (2009), 73, 84
Countryside, 19, 38–42
Critical theory, 24, 115, 116, 119

D
Dark Mountain project, 91, 100
Darwin, Charles, 36, 61, 62
Darwinism, *see* Social Darwinism
Deep adaptation, 95, 102, 108, 127
Depoliticization, 3, 73, 116, 120, 121, 123, 124, 127
Derrida, Jacques, 89, 90
Despair, 79, 85, 91, 96, 97, 100, 127
Dignity, 101–103, 106

E
Earth Summit (1992), 43
Ecological modernization, 22, 44, 69
Ecologist, The, 68
Ecology, *see* Nature interest, ecological
Ehrlich, Paul, 57–59, 61
Emissions trading, *see* Carbon trading
Emotions, v, vii, 6, 7, 18, 19, 23, 24n1, 54, 73, 79, 96–99, 101
Engels, Friedrich, 56

Enlightenment, 5, 8, 9, 30, 44, 72, 106, 117
Environmental movement, 6–7
Exploitation, 7, 20, 21, 30, 32n1, 36, 37, 48, 57, 69, 107, 114, 115, 119
Extinction Rebellion (XR), 72, 95–98, 101, 102, 106, 127
Extraction, 20, 44, 82, 83, 88, 121

F
Fanon, Frantz, 9
Fear, v, 3, 43, 49, 54, 55, 57, 58, 68, 71, 73, 74, 77, 79, 83, 102, 107, 108
Feuerbach, Ludwig, 118
Frankfurt School, *see* Critical theory
Franzen, Jonathan, 85
Fridays for Future, 72, 96, 97, 106, 127
Friends of the Earth, 67, 68
Fukushima, 84, 86, 125

G
Global Alliance for the Rights of Nature, *see* Rights of Nature
Globalization, 44
Global North, 3, 4, 19, 22–24, 31, 34, 38, 41, 44, 57, 71, 79, 80, 82, 84, 91, 92, 99, 107, 121, 124, 127
Global South, 3, 22, 23, 34, 44, 60, 61, 74, 80–82, 84, 85, 88, 98, 99, 107
Gore, Al, 3, 71, 84
Gramsci, Antonio, 116
Greene, Natalia, 89
Green New Deal, 5, 19, 45–47, 49, 122
Greenpeace, 3, 44, 64, 70

H
Habermas, Jürgen, 18
Hamilton, Clive, 2, 48, 84, 130
Hine, Dougald, 91–93, 100, 108
Hiroshima, 56
Hope, v, 3, 6, 9, 11, 24, 29–31, 43,
    48, 54, 62, 65, 73, 74, 79, 85,
    86, 91–93, 95–104, 106–108,
    115, 118, 119, 124, 126,
    127, 130
Horkheimer, Max, 48, 117
Hugo, Victor, 13

I
Ideology, 5, 7–12, 24, 38, 91,
    105, 115
Indigenous people, 32, 33, 35–37, 55,
    81, 82, 84, 86–88, 126
International Tribunal for the
    Rights of Nature, *see* Rights
    of Nature

J
Jacobs, Jane, 13
Jameson, Fredric, 10, 58, 116, 128
Jonstad, David, 108
Justice, 8, 15, 22, 42, 45–47, 54, 66,
    70, 74, 79–81, 85–91, 93, 101,
    104–107, 115, 116, 119,
    122–124, 126, 128, 129
    See also Climate justice

K
Kingsnorth, Paul, 91–94, 99,
    100, 108
Klein, Naomi, 6, 44–46, 82, 83, 120,
    126, 130

K
Kropotkin, Pyotr, 41, 59
Kyoto protocol, 22, 31, 42–44

L
Lenin, V.I., 35
Life reform movement, 21, 31, 38–42
Lukács, Georg, 13, 14, 17
Luxemburg, Rosa, 54, 71

M
Macy, Joanna, 103
Malm, Andreas, 17, 71, 108
Malthusians, *see* Neo-Malthusians
Mannheim, Karl, 105, 106
Marcuse, Herbert, 118, 130
Market mechanisms, 44, 84, 120
    See also Carbon trading
Marx, Karl, 8, 13, 14, 20, 61, 62, 114,
    115, 119
Matsumoto, Eiko, 77, 78
Melucci, Alberto, 2, 6, 11
Millennialism, 8, 105, 106
Mimesis, 117
Minamata Bay, 104
Montreal Protocol, 43
More, Thomas, 8
Mumford, Lewis, 48
Müntzer, Thomas, 54, 105

N
Naess, Arne, 70
Nagasaki, 56
Narrative, definition of
    apocalyptic narrative, 54
    green progress narrative, 29–30
    postapocalyptic narrative, 79–80
Natural history, 14, 16, 32, 36, 37, 55

Natural way of life, 39, 40
Nature, concepts of
  instrumental, 12, 13, 20, 30
  romantic, 12, 13, 16, 30, 32, 33, 39, 55, 93
  second nature, 13, 39
Nature interest
  definition of, 18–19
  aesthetic, 30, 32, 33, 35, 55, 118
  ecological, 19, 22, 31, 42, 45, 62, 64, 67, 69, 70, 88, 91, 92, 123
  economic, 22, 44, 121
  instrumental, 19–22, 30, 33, 36, 115
  juridical, 22, 87
  moral, 21, 39, 85, 93, 96, 101, 102, 126
  recreative, 34, 35, 55
Nazi movement, 41, 56
Neoliberalism, 42–45, 83
Neo-Malthusians, 57, 58, 61
Nuclear energy, 42

O
Oil, 80, 83, 84, 88
Osborn, Fairfield, 57, 58
Ozone layer, 43

P
Palme, Olof, 67, 68
Paris agreement, 43, 121
Paris COP meeting (2015), 73
Peak oil, 80, 84
Pinchot, Clifford, 33–35
Politicization, 3
Preservationists, 30, 32–38, 49, 56
Progress, *see* Narrative, definition of

R
Rancière, Jacques, 9, 106
Read, Rupert, 95–97, 102
Ricoeur, Paul, 10, 93, 118
Rights of Nature, 12, 22, 23, 80, 87, 89, 104, 105, 119, 126, 127
Roosevelt, Franklin D., 33, 38
Rosenberg, Alfred, 56
Ruong, Israel, 81

S
Sacrifice zone, 82, 83, 86, 88, 104, 105, 121
Sámi, 81, 82, 86
Säve, Pehr Arvid, 55
Schmidt, Alfred, 118
Schweitzer, Albert, 62
Scranton, Roy, 85
Second nature, *see* Nature, concepts of
Sierra Club, 33, 35
Smith, Neil, 14
Social Darwinism, 32, 33
Social Democrats, 43
Social movements, 5–12, 18, 21, 41, 54, 72, 99, 114
  See also Environmental movement, definition of
Soil Association, 67, 68
Solar energy, 43
Soper, Kate, 14
Standing Rock, 83, 103
Sustainable development, 24n1, 31, 42–45, 49
Swedish Society for Nature Conservation (SNF), 55

T
Tanaka, Shōzō, 78, 103, 126
Thunberg, Greta, 4, 72, 96, 98, 121
Transition movement, 3, 95, 96,
    99–101, 125

U
United Nations Framework
    Convention on Climate Change
    (UNFCCC), 43
Utopia, 5, 7–12, 24, 40, 48, 64,
    66–68, 72, 90, 93, 104–107,
    115–120, 123, 129, 130
Utsi, Inger, 81, 82
Utsi, Paulus, 81, 82

V
Vegetarian movement, 31, 40, 41
Vogt, William, 57–59

W
Warlenius, Rikard, 108
Wilderness, 15, 32–34, 37–39,
    93, 96
Wind energy, 36
World Wildlife Fund (WWF),
    3, 44

X
XR, see Extinction Rebellion